Practical Manual for Analysis of Soil, Water Fertilizer and Manure

THE AUTHORS

Dr. Javid Ahmad Sofi is Senior Scientist of Soil Science, Sher-e-Kashmir University of Agricultural Sciences & Technology of Kashmir (SKUAST-Kashmir). He has teaching experience of more than nine years and has taught at graduate and post-graduate level. At present he is working in Research Centre for Residue and Quality Analysis (RCRQA)- Referral Lab of the J&K state. He has expertise in ICP-OES, AAS, CHN analyzer, HPLC, GC, GC-MS etc. He received M.Sc. degree (2001) from SKUAST-K and Ph.D. degree (2012) from IARI, New Delhi. He cleared ICAR, NET (National Eligibility Test) conducted by Agricultural Scientists Recruitment Board (ASRB) of Indian Council of Agricultural Research (ICAR) in 2003. He has successfully completed externally funded ICAR project on Integrated nutrient management in Saffron. The author or co-author of over 40 research articles of national and international repute.

Dr. Shabir Ahmad Bangroo is Scientist of Soil Science, Sher-e-Kashmir University of Agricultural Sciences & Technology of Kashmir (SKUAST-Kashmir). He has also worked as Assistant Professor at Department of Soil Science and Agricultural Chemistry, Bihar Agricultural University, Sabour. He received M.Sc. degree (2010) and Ph.D. degree (2013) from SKUAST-K, Srinagar (J&K). He is the recipient of Science Talent Scholarship (2009) from J&K State Council of Science & Technology of India in 2009-2010 for PG programme and Maulana Azad National Fellowship of University Grants Commission (UGC) 2010-11 for PhD programme. He cleared ICAR, NET (National Eligibility Test) conducted by Agricultural Scientists' Recruitment Board (ASRB) of Indian Council of Agricultural Research (ICAR) in 2010. The author or co-author of over 30 articles of national and international repute.

Dr. Majeed Ul Hassan Chesti is Assistant Professor (Senior Scale) Soil Science at Directorate of Extension, SKUAST-Kashmir, Shalimar, Srinagar, J&K, India. He has completed his M.Sc. (Soil Science) in 2002 from Sam Higginbottom Institute of Agriculture, Technology and Sciences, Deemed University and PhD (Soil Science) in 2007 from SKUAST-Kashmir. He has been awarded Gold medal by securing first position in M.Sc. (Soil Science) as well as awarded merit certificate in PhD (Soil Science). He has more than 10 years of Research/Extension/Teaching experience and has also to his credit more than 35 research publications in peer reviewed journals and 5 book chapters. He has successfully completed two externally funded research projects sponsored by J&K State Council for Science and Technology and ICAR.

Practical Manual for Analysis of Soil, Water Fertilizer and Manure

–Authors–

Javid Ahmad Sofi

Shabir Ahmad Bangroo

Majeed Ul Hassan Chesti

2020

Daya Publishing House®

A Division of

Astral International Pvt. Ltd.

New Delhi – 110 002

Cataloging in Publication Data--DK
Courtesy: D.K. Agencies (P) Ltd. <docinfo@dkagencies.com>

Sofi, Javid Ahmad, author.
Practical manual for analysis of soil, water, fertilizer and manure / authors, Javid Ahmad Sofi, Shabir Ahmad Bangroo, Majeed Ul Hassan Chesti.
pages cm

ISBN 978-93-5130-999-4 (International Edition)

1. Soils--Analysis. 2. Water--Analysis. 3. Fertilizers--Analysis. 4. Manure--Analysis. 5. Plants--Nutrition. I. Bangroo, Shabir Ahmad, author. II. Chesti, Majeed Ul Hassan, author. III. Title.

S593.S64 2016 DDC 624.1514 23

Published by : **Daya Publishing House®**
A Division of
Astral International Pvt. Ltd.
– ISO 9001:2015 Certified Company –
4736/23, Ansari Road, Darya Ganj
New Delhi-110 002
Ph. 011-43549197, 23278134
E-mail: info@astralint.com
Website: www.astralint.com

Badrul Hassan
Director Education
SKUAST of Kashmir
Shalimar, Srinagar

Foreword

The Sher-e-Kashmir University of Agricultural Sciences & Technology of Kashmir is vested with the responsibility of promoting excellence in agricultural education in the state. There was a long felt need to provide comprehensive practical manual for analysis of soil , water , fertilizer and manure for the students, scientists and technicians working on this aspect. Javid Ahmad Sofi, Associate Professor (Soil Science) and his colleagues have expertise and recognition in the area of analytical techniques. He and his colleagues have brought out the present publication entitled as ***"Practical Manual for Analysis of Soil , Water , Fertilizer and Manure"*** which meets a long time felt need. I trust this publication will be of the great help to the students and instructors alike in the area of Soil Science.

This manual provides an exposition of the complex protocols through a simple, easy-to-understand step-by-step manner with good illustrations and examples. Aspects relating to soil testing, fertilizer adulteration and water analysis have been adequately covered and several practical's have been demonstrated. I am sure that this publication will be of immense use for researchers, teachers as well as students associated with soil science and crop nutrition. The authors deserve compliments for their in bringing out this useful manual.

Badrul Hassan

Preface

Appropriate knowledge about the analysis of soil, water, biological materials is vital in generating precise, accurate and dependable data for arriving at conclusive inferences. The need for the comprehensive and compact practical manual on soil, water, fertilizer and manure analysis was felt ever since it was recognized as a compulsory course in soil science and agronomy. The publications emanating from our country have focused on either physical or chemical or biological properties of soil. The present publication meets the requirement of the practical components covering all aspects of soil science in B.Sc and M.Sc Agriculture in its endeavour to expose students in the modern and the latest techniques to ensure quality education in the country.

For the first time a compact practical manual has been developed, which includes practical exercise not only relating to physical or chemical aspects but also practical's hitherto not covered like manure analysis and fertilizer adulteration tests. These exercises have been grouped into ten chapters. First three chapters have been devoted to give elementary idea of about the role soil in agriculture and can be very useful for students and academicians working at different institutions. Other seven chapters give the detailed methodology of soil, water, fertilizer and manure analysis. Calculations have been also given with examples to have better understanding of the subject. No effort has been spared in presenting the information in this publication in a clear and concise manner and is suitably illustrated.

Authors gratefully expressing their thanks to colleagues and students who have contributed to prepare this manual. Thanks are due to scientific staff of Division of Soil Science, SKUAST-Kashmir, for their keen interest and boosting support for this publication. It is hoped that this publication will be useful not only to the students but also to the scientists working on soil science and crop nutrition at state agricultural universities (SAU's) as well as to other personal engaged in soil testing laboratories. Authors look forward to receive suggestions for improvement of the manual.

Javid Ahmad Sofi
Shabir Ahmad Bangroo
Majeed Ul Hassan Chesti

Contents

CHAPTER-1

Soils

Soil may be defined as a thin layer of earth's crust which serves as a natural medium for the growth of plants. It is the unconsolidated mineral matter that has been subjected to and influenced by genetic and environmental factors – parent material, climate, organisms and topography all acting over a period of time. Soil differs from the parent material in the morphological, physical, chemical and biological properties. Also, soils differ among themselves in some or all the properties, depending on the differences in the genetic and environmental factors. Thus some soils are red, some are black; some are deep and some are shallow; some are coarse-textured and some are fine-textured. They serve in varying degree as a reservoir of nutrients and water for crops, provide mechanical anchorage and favorable tilth. The components of soils are mineral material, organic matter, water and air, the proportions of which vary and which together form a system for plant growth.A developed soil will have a well defined profile which is a vertical section of the soil through all its horizons and it extends up to the parent materials. The horizons (layers) in the soil profile which may vary in thickness may be distinguished from morphological characteristics which include colour, texture, structure etc. Generally, the profile consists of three mineral horizons – A, B and C. The A horizon may consist of sub-horizons richer in organic matter intricately mixed with mineral matter. Horizon B is below A and shows dominance of clay, iron, aluminium and humus alone or in combination. The C horizon excludes the bedrock from which A and B horizon are presumed to have been formed. A study of the soil profile is important from crop husbandry point of view, since it reveals the surface and the sub-surface characteristic and qualities namely, depth, texture, structure, drainage conditions and soil moisture relationship which directly affect the plant growth. *A study of soil profile supplemented by physical, chemical and biological properties of the soil will give full picture of soil fertility and productivity.*

1.1 Soil Composition

a) **Solid Phase:** It includes minerals (sand, silt and clay) and organic matter both dead and living micro and macro-organisms.

b) **Liquid Phase:** Soil water which occupies the pore spaces (between mineral particles) carries the plant nutrients to the roots.

c) **Gasous Phase:** Also called as soil air, occupies pore spaces similar to soil water. It fills these voids when soil water is absent. It carries respiratory products of roots and soil-organisms. It has a higher concentration of carbon dioxide than atmospheric air.

Soils in nature have different properties such as physical properties which include, soil texture, soil strength, soil structure, bulk density, soil aeration, soil water, soil temperature, infiltration capacity, hydraulic conductivity. Chemical properties depending on mineralogical composition and type inorganic soil colloidal fraction, include, pH, CEC, organic matter etc. The biological properties include extent and type of bacteria, fungi, actinomycetes, algae, earthworms, soil enzymes etc.

1.2 Physical Properties

a) Soil texture: Refers to the relative proportion of sand, silt and clay present in a soil. Based on these proportions soils are classified into various textural classes. Clayey soils have a larger percent of clay. They are considered more fertile than sandy soils but are difficult to work. Sandy soils are easy to work but are less fertile. They have low water retention capacity. Loamy soils are in between sandy and clayey soils. They are best for arable cropping.

b) Soil structure: Refers to the arrangement of the different particles into soil aggregates. Roots move between these aggregates. A compact soil will resist root movement. The organic matter content helps the soil aggregation process.

c) Bulk density: It is the weight of the soil in the given volume. A compact soil has a higher value while an organic soil has a lower value. It also affects water holding capacity of the soil.

d) Field capacity: Refers to the moisture content of a soil after the loss of gravitational water. At this point, water is held in soil micro-pores, which is available to plant roots, until the water content down to a lower value. This lower value is referred to as the permanent wilting point. The amount of water available between field capacity and permanent wilting point is referred to as available water. This value is important in determining irrigation.

1.3 Chemical Properties:

a) Soil pH: Measures the negative logarithm of the hydrogen ion activity of the soil solution. It is a measure of the soil acidity or alkalinity of a soil.

b) Soil acidity: It is caused by many factors such as excessive rain which leaches basic cations (Ca, Mg, K), use of nitrogenous fertilizer like urea, ammonium sulphate etc., and oxidation of iron pyrite containing minerals.

c) ***Salinity and Alkalinity:*** Occurs in arid and semi arid regions, where precipitation is insufficient to meet evapo- transpiration needs of plants, when salts move up to the surface. Salt affected soils occur within irrigated lands. Salts are added through irrigation water through over-irrigation and salts accumulate in poorly drained areas. Salt content in soils is measured in terms of electrical conductivity (EC). A saline soil has a pH of less than 8.5, but soils are well flocculated. The EC is more than 4 deci Seimens per meter (4 dS m^{-1}). Soils which are saline cause problems to crops during dry weather. Loss of crop yield from poor growth is common. Alkaline soils have high pH (more than 8.5) and a high concentration of sodium in them. Alkaline soils are deflocculated, drainage is poor and growing plants is difficult due to high pH and higher content of sodium in the soil. Draining, flushing after ploughing and addition of organic matter and gypsum could correct these problems.

d) ***Cation exchange capacity***: The power to retain cations at the surface of soil colloids is referred to as the cation exchange capacity. Soil colloids of clay and organic matter have this property due to the presence of negative charges at the surface. CEC is defined as the sum of cations held by a kilogram of soil. It is expressed in $Cmol_c Kg^{-1}$ soil. Clays like kaolinite have low CEC and the CEC is pH dependent. Organic matter has a large CEC but it is too pH dependent. Montmorillinitie clay has high CEC due to the negative charges developed through loss of cations during formation of these clays. Practices like fertilization, liming, irrigation and addition of organic manures can increase the exchangeable cations.

e) ***Soil organic matter***: Consists of living organisms, dead plant and animal residues. It is the most chemically active portion of the soil. It is a reservoir for various essential elements. It contributes to CEC, promotes good soil structure, buffers soil pH and promotes good air and water relations in plants.

1.4 Biological Property

An important function of microbes on this planet is to facilitate the recycling of elements. The organically –bound nutrient elements are brought back to their mineral forms by soil organisms.

a) ***Soil enzymes-***: Enzymes are the vital activators in life processes, likewise in the soil they are known to play a substantial role in maintaining soil health and its environment. The enzymatic activity in the soil is mainly of microbial origin, being derived from intracellular, cell-associated or free enzymes. They are important in catalyzing several vital reactions necessary for the life processes of micro-organisms in soils and the stabilization of soil structure, the decomposition of organic wastes, organic matter formation, and nutrient cycling, hence playing an important role in agriculture.

b) ***Mineralization***: Conversion of N in organic residues and soil organic N into soluble forms occurs through mineralization. Carbon sources are degraded sources of energy, N in excess of microbial need is liberated. The following sequence of reactions takes place.

   i) ***Ammonification***: Complex protein compounds are broken down to ammonium compounds by micro-organisms.

   ii) ***Nitrification***: Ammonium compounds are oxidized to nitrite and to nitrate by two specific types of soil bacteria, *Nitrosomonas* and *Nitrobacter*.

   iii) ***Denitrification***: The nitrates are reduced to nitrogen gas under poorly aerated conditions through specific micro-organisms.

CHAPTER-2

Plant Nutrient and their Functions

Several elements take part in the growth and development of plants, and those absorbed from the soil are generally known as plant nutrients. Besides these, the plant takes up carbon, oxygen and hydrogen, either from the air or from the water absorbed by roots. In all, 17 elements have been identified and are established to be essential for plant growth. These are carbon (C), hydrogen (H), Oxygen (O), nitrogen (N), phosphorus(P), potassium(K), calcium (Ca), magnesium (Mg), sulphur(S), iron (Fe), zinc (Zn), manganese (Mn), copper (Cu), boron (B), molybdenum (Mo), chlorine (Cl) and Nickel (Ni). These elements serve as raw materials for growth and development of plants, and formation of fruits and seeds. In the soil a nutrient element is found in abundant quantities, however these nutrients are tied up in mineral and chemical compounds. The roots cannot absorb them for synthesis of various plant constituents and hence need for assessing the plant available amounts of nutrients in the soil and meeting deficiency by application of manures and fertilizers to such soils for optimum crop production.

2.1 Plant Nutrients

An essential nutrient element is one that is required to complete the life cycle of the organism and its relative deficiency produces specific deficiency symptoms. The adverse effects are relieved by the supply of that specific element only. Some of these are required in large amounts and some in traces. These are classified as major and micro nutrients, and are further classified as follow:

2.2 Major Nutrients or Macro-nutrients

The macro-nutrients (C, H, O, N, P, K, Ca, Mg and S) are those found in comparatively high concentrations in plants. The C, H and O constitute 90-95 percent of the plant dry matter by weight and are supplied through CO_2 and water. Remaining six major nutrients are further subdivided into primary and secondary nutrients.

2.2.1 Primary Nutrients

Nitrogen, P and K are termed primary nutrients because of their larger requirement by the plant and correction of their wide spread deficiencies are often necessary through application of commercial fertilizers of which these are the major constituents.

2.2.2 Secondary Nutrients

Calcium, Mg and S are termed as secondary nutrients because of their moderate requirements by plants, localized deficiencies and their advertent alleviation by incidental accretion through carriers of primary nutrients.

2.2.3 Micro-nutrients

Nutrients that are required in relatively smaller quantities but are as essential as macronutrients are termed as micronutrients. These include iron, manganese, zinc, copper, boron, chlorine and nickel.

It has been found that the presence of some elements which are not considered essential for plant growth and are not directly concerned in the nutrition of the crop, but are present in the plants used as food and feed, are of vital importance to the health of human beings and animals. The elements within this group are iodine, cobalt and sodium. In addition, there is another group of elements which are toxic to the animals feeding on the plants containing them. These are selenium, lead, thallium, arsenic and fluorine. Elements, such as sodium, fluorine, nickel, lead, arsenic, selenium, aluminium and chromium, when occurring in soils in high available amounts, may also prove toxic to the plants and restrict their growth. Some elements, occurring freely in the soil, are absorbed by the plants as impurities. They may occasionally stimulate growth although they are not essential for plant growth. They include lithium, strontium, tin, radium, beryllium, vanadium, barium, mercury, silver and bromine. Silica is reported to be 'beneficial', particularly in rice but is not classified as essential as per the criteria fixed for this purpose.

Plant nutrients are usually absorbed through roots. Roots have the ability to absorb nutrients selectively. Nutrients reach to the plants roots by mass flow, diffusion and root interception.

Root absorption takes place both as active and passive absorption.

- Active absorption takes place as an exchange phenomenon and requires energy. Most plant nutrients are absorbed in this manner.
- Passive absorption is part of the transpiration cycle (mass flow). Water and some dissolved solutes are absorbed by this process.

2.3 Deficiency of an Element

Nutrient deficiency symptoms usually appear on the plant when one or more nutrients are in short supply. In many cases, deficiency may occur because an added nutrient is not in the form the plant can use. Deficiency symptoms for specific elements are included on the "Key to Nutrient Disorders". Nutrient deficiency results into: (i) Complete crop failure at the seedling stage (ii) Severe stunting of plants

(iii) Delayed or abnormal maturity (iv) Obvious yield differences (v) Poor quality of crops, including differences in protein, oil or starch, sugar content and storage quality. Visual Deficiency symptoms are excellent diagnostic aids for detecting nutrient deficiencies, but have limitations

- ***Deficiencies caused by more than one nutrient:*** It is not easy to relate a deficiency symptom to the deficiency of a particular nutrient (e.g. Yellowing of the leaf or chlorosis is shown with deficiency of more than one nutrient). A trained eye or history of soil management can arrive at a good diagnosis.

Table 2.1. Nutrients Essential for plant growth and the forms in which taken up by the plants

Nutrients	Chemical symbol	Form taken up by the plant
Primary nutrients		
1. Carbon	C	CO_2, HCO_3^-
2. Hydrogen	H	H_2O
3. Oxygen	O	H_2O, O_2
4. Nitrogen	N	NH_4^+, NO_3^-
5. Phosphorus	P	$H_2PO_4^-$, HPO_4^{2-}
6. Potassium	K	K^+
Secondary nutrients		
7. Calcium	Ca	Ca^{2+}
8. Magnesium	Mg	Mg^{2+}
9. Sulphur	S	SO_4^{2-}
Micro-nutrients		
10. Iron	Fe	Fe^{2+}, Fe^{3+}, Chelate
11. Zinc	Zn	Zn^{2+},$Zn(OH)_2$, Chelate
12. Manganese	Mn	Mn^{2+},Chelate
13. Copper	Cu	Cu^{2+},Chelate
14. Boron	B	$B(OH)_3$
15. Molybednum	Mo	MoO_4^-
16.Chlorine	Cl	Cl^-
17. Nickel	Ni	Ni^{2+}

- ***Deficiencies are actually relative:*** Deficiency of one nutrient may be related to an excessive quantity of another (e.g Mn deficiency may be induced by adding large quantities of Fe).
- ***Deficiencies difficult to distinguish:*** A disease or insect damage may resemble certain nutrient deficiencies (leaf hopper damage can be confused with B deficiency in alfalfa).
- ***Deficiencies caused by more than one Factor:*** Sugars in maize combine with flavines to form anthocyanins (purple, red, and yellow pigments) and their accumulation may be caused by an insufficient supply of P, low temperature, insect damage to the roots or N deficiency.

Insufficient, sufficient and toxic levels: Nutrient contents associated with only growth reduction and not accompanied by appearance of deficiency symptom are termed as insufficient. Range of nutrient content in plants associated with optimum crop yields is called sufficient. When the concentration of the nutrients rise too high to cause significant growth reduction. It is termed as toxic.

2.4 Fate of Nutrient Elements in Soil

Nutrients are lost through *Crop removal, Erosion, Leaching, Volatilization, Denitrification and Fixation.*

2.4.1 Crop Removal

Plant species have specific requirements of plant nutrients. Nutrient removal depends on growth condition, crop sanitation, cultivation and yield obtained. Grain crops require more nitrogen than other nutrients. Pulses require more phosphorous while crops such as potato, apple, banana and pineapple require more potassium compared to other nutrients.

2.4.2 Erosion

Entire top soil is lost through erosion by water or wind, these results in loss of soil nutrients.

2.4.3 Leaching

Water percolating through a soil profile carries dissolved nutrient elements. Nutrients are easily lost in humid regions and sandy soils. Bare soil loses more nutrients than cultivated soils.

2.4.4 Volatilization

Nitrogen is easily lost through volatilization as ammonia, particularly in paddy soils and upland soils in poorly drained areas. This is referred to as ammonia volatilization. This loss is enhanced by high temperature and wind.

2.4.5 Denitrification

Nitrate form of N is lost through denitrification where nitrogen gas or nitrous oxide is released. This loss occurs mainly in paddy soils and in upland soils which are saturated with water periodically or part of the season.

2.4.6 Fixation

Fixation takes place by conversion of a nutrient to an unavailable form. Phosphorus is converted to unavailable forms both in acidic and alkaline soils, as aluminium/iron phosphate or tricalcium phosphate respectively. Potassium and ammonium N can be fixed by certain clay minerals.

2.5 Function of Nutrients in Crop Productions.

In the nutrition of a crop various nutrients perform distinct functions. Their relative essential role depends upon whether they enter into chemical composition or regulate the various physiological processes in the plant.

2.5.1 Carbon

The plant absorbs carbon dioxide directly from the atmosphere. This combines with water in the presence of light and forms the primary sugars, such as glucose and fructose (fruit sugar). Chlorophyll is the pigment which absorbs the radiant energy of the sun and brings about complex chemical synthesis of carbon dioxide and water resulting in simple sugars. This process is called photosynthesis or synthesis in light. Thus, by the combination of carbon with water, sun's energy is stored in the plant body, and the first carbohydrates are formed in the plant. From the carbohydrates complex sugars, starches, hemicelluloses and celluloses are formed. These simple sugars also polymerase (chemically combine) into oils and fats. The process of respiration degrades organic compounds to provide energy for various plant metabolic processes.

2.5.2 Hydrogen

It is one of the most important elements in the nature. It readily combines with oxygen to form water and with carbon to form complex chemical organic compounds. The growth of plants would only take place if adequate quantity of water is supplied to meet the needs of hydrogen for synthesis of organic substances. When organic compounds either breakup in the plant or decompose in the soil or atmosphere, the released hydrogen always combines with oxygen and forms water. Thus, the exchange of hydrogen takes place in either of the synthesis or decomposition (including respiration) processes. Hydrogen ions are involved in electrochemical reactions and maintain electrical charge balances across all membranes.

2.5.3 Oxygen

Oxygen is part of water as well as carbon dioxide. When water combines with carbon dioxide oxygen is evolved:

$$6CO_2 + 6H_2O \longrightarrow C_6H_{12}O_6 + 6O_2$$

Thus, the oxygen evolved in the process equals the volume of carbon dioxide absorbed by the plants. This evolution of oxygen takes place in the process of photosynthesis. The process is reverse in respiration when a simple or a complex sugar, fat or oils breaks up. It requires oxygen and gives out CO_2.

$$C_6H_{12}O_6 + 6O_2 \longrightarrow 6CO_2 + 6H_2O \text{ (Glucose)}$$

Six molecules of carbon in the glucose combine with six molecules of oxygen to form six molecules of carbon dioxide. In the process six molecules of water are formed. Thus, oxygen plays a dominant role in the processes of photosynthesis and respiration in plants. Oxygen is necessary for all oxygen-requiring reactions in plants including nutrient uptake by roots.

2.5.4 Nitrogen

It is an essential component of amino-acids, protein, nucleic acids, porphyrins, flavins, purines and pyrimidine, nucleotides, flavin nucleotides, enzymes, co-enzymes and alkaloids. Being a constituent of nucleic acids viz. Ribo Nucleic Acid

(RNA) and Deoxyribo Nucleic Acid (DNA), nitrogen is responsible for the transfer of genetic code to the off-springs. Nitrogen containing chlorophylls in the presence of solar energy fixes atmospheric CO_2 as carbohydrates. It improves quality of leafy vegetables and fodders. Nitrogen fertilization improves protein quality of the food-grain by enhancing the proportion of glutamic acid, proline, phenylalaline, cystine, methionine, and tyrosin and decreasing the amount of lysine, histidine, arginine, aspartic acid, threonine, glycine, valine and leucine in the grain.

2.5.5 Phosphorus

Once inside the plants, P gets incorporated into nucleic acid, phospho-proteins, phospholipids, sugar phosphates, enzymes ATP (adenosine triphosphate) and ADP (adenosine diphosphate).The ATP and ADP are the energy currency of the plants. Major processes involving ATP are generation of membrane electrical potentials, respiration, biosynthesis of cellulose, hemicelluloses, pectins, lignins, proteins, lipids, phospholipids and nucleic acids. It is used in several energy transfer compounds in plants Phosphorus is particularly helpful in the production of legumes, as it increases the activity of nodular bacteria which fix nitrogen in the soil. It aids the formation of seeds and fruits, particularly in the legumes. It stimulates early root growth and development. Optimum quantity of phosphorus available to the crop in combination with nitrogen balances their shoot and root growth.

2.5.6 Potassium

Potassium plays a major role as an activator in many enzymatic reactions in the plant. Many enzymes responsible for cellular reactions require K as a co-factor. It regulates the opening and closing of stomata, which are essential for the photosynthesis, water and nutrient transport, and plant cooling. It plays a major role in transport of water and nutrients throughout the plant in xylem. Under reduced K supply translocation of NO_3^-, PO_4^{3-}, Ca^{2+}, Mg^{2+} and amino-acids are hampered. It increases root growth and improves drought tolerance. It activates a large number of enzymes (more than 60), by way of exposing the active reaction sites. Potassium is responsible for the activation and synthesis of protein forming nitrate reductase enzyme. Under high K levels starch moves efficiently from sites of production to storage. Potassium enhances the crop quality, shelf –life and vegetables. The stalks and stems (of plants) are more stiff when an adequate supply is available than otherwise. In consequence the lodging in cereals is reduced. It increases the plumpness of the grains. In general it imparts vigour and resistance to diseases. Some crops, such as potato, tomato, clovers, lucerne and beans, are more responsive to potassium than other crops. As larger quantities of carbohydrates and proteins are stored, it increases winter hardiness of some plants, such as lucerne.

2.5.7 Sulphur

Sulphur is a constituent of many proteins, and aids in the formation of chlorophyll and root growth. It is a component of sulfur - containing amino acids such as cysteine, cystine, methionine. Plants having an abundant supply of sulphur develop dark-green leaves and extensive root system. In legumes the nodular activity is appreciably increased by adequate supply of sulphur, because it is a constituent of ferrodoxin-containing nitrogenase, which takes part in the biological nitrogen fixation. It plays a major role in increasing the oil quality in oilseed crops. Due to larger availability of proteins the plant growth is vigorous. It improves the starch content of tubers.

2.5.8 Calcium

It is constituent of calcium pectate in the cell wall and maintains the integrity of membrane. It is important for the growth of meristems and functioning of root tips. It plays a role in cell division and helps maintain the chromosome structure. It also helps to keep up sustained activity of the nodule bacteria in legumes. Besides its direct nutrient value, calcium when applied to acid soils increases the availability of other nutrients, like phosphorus, nitrogen and molybdenum. Excess of calcium in the calcareous soils depresses the uptake of potassium and magnesium. These are secondary effects of calcium on plant growth.

2.5.9 Magnesium

It is important constituent of chlorophyll which is indispensible for photosynthesis. It maintains the dark-green colour of leaves and regulates the uptake of other materials, particularly nitrogen and phosphorus. It appears to play an important role in the transport of phosphorus, particularly into the seeds. It is an activator of many enzyme systems involved in carbohydrate metabolism and synthesis of nucleic acids.

2.5.10 Iron

Iron is used in the biochemical reactions that form chlorophyll and is a part of one of the enzymes that is responsible for the reduction of nitrate-N to ammonical -N. Other enzyme systems such as catalase and peroxidase also require iron. Although iron does not enter into the composition of chlorophyll, its deficiency manifests itself in *chlorosis, yellowing* or *whitening of leaves*. The concentration of iron ions plays an important part in the oxidation process in leaf cells. When iron is not taken up in adequate quantity, the growth of plants is less vigorous, and seed and fruit development suffer as a consequence of decreased photosynthetic activity in the leaves. Too much of liming results in iron deficiency. Sever deficiency results in chlorosis and leaves turn white and eventual leaf loss.

2.5.11 Zinc

In a general way zinc is associated with the development of chlorophyll in leaves and a high content of zinc is correlated with a high amount of chlorophyll. In its absence growth is less, buds fall off and seed development is limited. Zinc is involved in the synthesis of indole acetic acid (IAA), metabolism of gibberellic acid and synthesis of RNA. Zinc influences translocation and transport of P in plants. Under Zn-deficiency, excessive translocation of P occurs resulting in P-toxicity. In peach and apricots, zinc deficiency symptoms are manifested in leaf. In small trees bronzing of leaves is mitigated by spraying zinc sulphate on leaves. The citrus mottling of leaves may be frequently due to the deficiency of zinc in the plant.

2.5.12 Manganese

Manganese also activates several enzymes and is involved in the processes of the electron transport system in photosynthesis. Manganese is an essential element and appears to have a role in the formation or synthesis of chlorophyll. Due to deficiency of manganese the carbohydrate synthesis is disturbed, resulting in retarded growth, decrease in the content of ash and failure to reproduce. The leaves and roots of plants deficient in manganese have much less of sugars than those which can absorb sufficient quantity of manganese. Manganese, probably in association with iron, is a constituent of some respiratory enzymes and some enzymes responsible for protein synthesis from the amino acids formed in the leaves.

2.5.13 Copper

In the chloroplasts of leaves there is an enzyme which is concerned with the oxidation-reduction processes. The presence of copper is essential for this enzyme to function. Thus, copper plays an important role in the process of photosynthesis. Copper is important in imparting disease resistance to the plants. It enhances the fertility of male flowers.

2.5.14 Boron

Boron is responsible for the cell wall formation and stabilization, lignifications and xylem differentiation. It plays a role in pollen germination and pollen tube growth. It facilitates ion uptake by way of increasing the activities of plasma-membrane bound H^+-ATPase. Boron imports drought tolerance to the crops. It also facilitates transport of K^+ in guard cells as well as stomatal opening. Boron deficiency affects the young growing points of stem, buds, leaf tips and margins, and root tips. Buds develop necrotic areas and leaf tips become chlorotic and eventually die. One of the most marked effects of boron deficiency observed is the restricted development of nodules on the roots of legumes. Very little nitrogen is fixed in these nodules. The tissues carrying the minerals and water from the soil to the leaves are also disorganized in the absence of boron. Cole crops, beets, and celery have rather high B requirements, otherwise only small amounts of B are needed by plants and supplying excessive B from fertilizer or from foliar sprays can lead to toxicity.

2.5.15 Molybdenum

Biological nitrogen fixation is catalysed by the Mo-containing enzyme, nitrogenase. Thus presence of molybdenum is very essential for the fixation of atmospheric nitrogen in the roots of legumes by nodule bacteria. Nitrate is reduced by nitrate reductase (NR) enzyme in cytoplasm by transfer of electrons from Mo to NO^-_3 .Owing to close relationship between Mo supply, nitrate reductase activity (NRA) and plant growth, NRA has been used as an indicator of status of Mo in plants. Molybdenum affects the formation and viability of pollen and development of an anthers..

2.5.16 Nickel

Nickel is metal component of urease that catalyse the reaction $CO\ (NH_2)_2 + H_2O = 2NH_3$. Apparently, Ni is essential for plants supplied with urea and for those in which ureides are important in N-metabolism. Ni- deficient plants accumulate toxic levels of urea in leaf tips because of reduced urease activity. Ni-deficient plants may develop chlorosis in the youngest leaves that progresses to necrosis of the meristem. In free-living *Rhizobia,* adequate Ni supply ensures optimum hydrogenase activity. Nickel facilitates transport of nutrients to the seeds or grains.

2.5.17 Chlorine

It plays a major role in osmoregulation (Cell elongation, stomatal opening) and charge compensation in higher plants. Chloride is involved in photosynthesis and functions as a counter-ion in maintaining turgor pressure in cells. Chlorine improves the nutritional quality of vegetables. The chloride ion is very common in the environment and is often found as a constituent in fertilizers; therefore, deficiency symptoms are rare.

CHAPTER-3

Deficiency Symptoms of Nutrients in Plants

Approximately 80% of all nutrients absorbed by roots are translocated to the shoots. When nutrient supply is abundant, they are delivered directly to the shoots often within minutes of absorption. Accordingly, plants may absorb and accumulate essential elements in far greater quantities than are necessary for immediate use. These accumulated elements are available for use later in the plant life cycle when demands are high for fruit production and/or when nutrient supply from the soil is restricted. The ability of an element to move from one plant part to another is called mobility and the process is known as re translocation. The mobility of the essential elements in plants is given below:

Highly mobile (N, P, K, and Mg)

Moderately mobile (Zn)

Less mobile (Fe, Mn, Cu, Mo and Cl)

Immobile (Ca and Boron)

The mobility of an element influences the location where deficiency symptoms are likely to be observed on the plant. For example, Mg deficiency symptoms occur on the oldest, generally lower leaves, because Mg is retranslocated to the younger leaves of the plant. Conversely, Ca deficiencies occur at the growing point or in storage organs like roots and fruits because Ca, being immobile, is not retranslocated to these sites during Ca stress conditions.

The plants exhibit deficiency symptoms that are characteristic for each element, and are, therefore useful for diagnostic purposes. However, in many cases, the symptoms may be masked by symptoms of other nutritional disorders, those caused by unfavorable environment, or stress caused by plant pests. In these situations, plant tissue analysis provides useful information to complement and confirm visual diagnosis. The general symptoms associated with deficiencies of the essential elements follow:

3.1 Nitrogen

Nitrogen deficiency first appear on older leaves in the form of light green to pale yellow colouration. Under excessive nitrogen, the succulence of the plants increases, taller plants and heavier heads causes lodging, more susceptible to insect, pest and disease attacks. In fruit trees nitrogen deficiency appears as a general light green color or yellowing of the lower leaves, in peaches and nectarines, the leaves also appear reddish on the margins. Very low N trees have stunted growth, poor fruit set, and smaller size, compared to high N trees. In citrus the leaf shedding is heavy. Their leaves are small in size, thin and fragile and have light green colour. In deciduous fruit trees the leaves have yellowish green appearance. The old-mature leaves are discoloured from base to tip. Under prolonged deficiency twigs become hard and slender. In small grains, namely, wheat, barley and oats, the nitrogen-starved plants are erect and spindly and the leaves have yellowish-green to yellow colour. The stems are purplish-green. In corn if nitrogen deficiency persists the yellowing will follow up the leaf mid rib in the typical V-shaped pattern with the leaf margins remaining green. The drying up of lower leaves is generally referred to as firing. In potato, in the later stages of growth, the margins of lower leaflets lose their green colour and become pale-yellow. In cotton the blades and petioles are reduced in size, turn yellow or brown and die. Plants produce fewer lateral branches, reduced number of fruiting branches, and very much reduced number of flowers and bolls. In legumes the growth is stunted and the lower leaves are pale-yellow or brownish in colour. In vegetables there is retarded growth with leaf chlorosis. The stems are slender, fibrous and hard.

3.2 Phosphorus

Because of its faster mobility in plants, it gets easily translocated from older tissues to the meristematic tissues. Therefore, deficiency symptoms of P appear first on older leaves. Phosphorus deficiency, being associated with the accumulation of carbohydrates, results in the production of dark green colour leaves. In fruit trees leaves are small but dark green and they develop autumn tints early, particularly deep reds. The older leaves of peach trees develop purple spots on margins which roll upwards. Flowering is often sparse. Generally the plant is dark-green but the lower leaves may turn yellow and dry up. In citrus, the plants show reduced growth and the older leaves at first lose their deep-green colour and luster, and develop faded green to bronze colour. In deciduous fruit trees the young leaves have dark-green colour while mature ones have bronze or ochre dark-green colour. In corn, leaves and stems have a tendency to become purplish, young plants are stunted and dark-green in colour. Small grains have dark-green colour and often have purplish tinge. They have retarded growth. In potato, in early stages, the plants have stunted spindly growth. The tubers have rusty-brown lesions in the flesh in the form of isolated flecks which sometimes join together to produce larger discoloured areas. The cotton plants have dark-green colour, leaves and stems are small, and the bolls mature late. Besides the dark-green colour of legume plants their petioles and leaflets are tilted upwards. And their stems often turn red. The new twigs are slender. In vegetables although the growth is retarded the leaves do not show symptoms of chlorosis. In many crops the under surface of leaves develops reddish-purple colour. The stems are slender and woody. They bear small, dark-green leaves.

3.3 Potassium

Potassium deficiency symptoms develop first appear on the older leaves. In fruit trees when potassium is deficient, leaves are pale green -yellow and tend to curl inward (boating) and burn along the margins and tips. Poor fruit size is commonly associated with low potassium. In deciduous trees the necrosis (death of tissues) in foliage occurs, the necrotic areas varying in size from very small dots to patches or extensive marginal areas. Foliage, especially of peach, becomes usually crinkled. In maize, in the young stage, the edges and tips become dry and appear scorched or fired. At a later stage in well-grown plants the leaves are streaked with yellow and yellowish-green colour, and the margins dry up and get scorched. Similar symptoms are shown by oats, wheat and barley. In potato the deficiency of potassium is acutely manifested. The plant growth is retarded, the internodes are somewhat shortened, the leaf size is reduced and they form a sharper angle with the leaf petiole. The leaflets become crinkled and curve downward. The older leaves become yellowish, develop a brown or bronze colour, starting from the tip and edge and gradually affecting the entire leaf, and finally die. In legumes the first symptoms consist of yellow mottling around the edges of the leaf. In vegetable crops in the older leaves bronze and yellowish-brown colours are manifested near the margins. Specks develop along the veins of the leaf and ultimately the tissue deteriorates and dies.

3.4 Sulphur

Sulphur deficiency first appear on the younger growths as it is immobile in the plants. The fading of the normal green colour of the young leaves, followed by chlorosis is the most common deficiency symptoms. The growth of citrus slows down and the younger leaves develop very light yellow-green to yellow colour. In *Brassicas*, which are more susceptible to S-deficiency, leaves show cupping owing to the curling of the leaf margins and arresting of the growing points. The older leaves become puckered with inward raised areas between the veins. In vegetables, the leaves develop yellowish-green colour, and become thick and firm. The stems harden and sometimes become abnormally elongated and spindly. In legumes the younger leaves turn pale-green to yellow, while terminal buds remain alive.

3.5 Calcium

Calcium is immobile in the plant and cannot be readily tranlocated from the older to the new leaves and hence deficiency symptoms appear on the younger structures. Generally the deficiency symptoms due to calcium starvation are localized in new leaves and in bud leaves of plants. Under Ca-deficient, root system is stunted. In severe cases the terminal bud dies. In citrus the green colour fades along the edges of the leaf and this spreads to areas between veins. The symptoms appear first in immature leaves of deciduous fruit trees, especially those at the top which dieback from tips and margins or along the midribs. In maize the tips of the unfolding leave gelatinize and when they dry they stick together. In potato a light green band appears along the margins of the young leaves of the bud and the leaves often have a wrinkled appearance. In vegetables the stems grow thick and woody, and the new leaves are chlorotic and lack turgidity. In legumes the nodules developed are small and fewer in number. *Blossom end rot* in tomatoes and *Bitter pit* in apples are the Ca-deficiency symptoms.

3.6 Magnesium

Magnesium is mobile within the plants with its symptoms of deficiency appearing in terms of interveinal chlorosis and streaked or patchy effects on older leaves. In fruit trees this element when deficient appears as marginal chlorosis on the leaves in an inverted "V" shaped pattern. It mostly affects basal leaves of the shoot while terminal leaves are not affected. It usually occurs where there is very high potassium (K) in the soils and on young trees. *Grass tetany* is a nutritional disorder common in cattle grazing on Mg-deficient pastures well fertilized with potassium. In deciduous fruit trees necrosis occurs as fawn-coloured patches on most mature, large leaves. The affected leaves drop, leaving a tuft or rosette of thin, dark-green leaves at the terminal part of the twigs. In small grains the plants are dwarfed and turn yellow. Sometimes leaves exhibit yellowish-green patches. In maize leaves a slight yellow streak develops between the parallel veins in the leaves. In acute deficiency these streaked tissues may dry up and die. In potato the affected leaves are brittle. In legumes the areas between main veins of the leaves become pale-green, which later turn deep yellow. At a later stage of growth the leaf margins curl downwards accompanied by a gradual yellowing and bronzing from the margin inward. The chlorosis in legumes begins at the tip and margins of the lower most leaf, and progresses between the veins towards the centre of the leaflet. Eventually the tissue between the veins is filled with brown, dead areas. In vegetable crops the symptoms are similar. The chlorosis appears first between leaf veins of new leaves and then spreads to older leaves. The chlorotic areas become brown or transparent and ultimately marked necrosis of affected tissue occurs.

3.7 Iron

Deficiency of iron results in interveinal chlorosis appearing first on the younger leaves with leaf margins and veins remaining green. Bleached yellow to white leaves with green veins are the symptoms most often seen. Leaf size is usually normal. The fruit tends to ripen early and quality is normal. If the tree is severely deficient, causing shoot dieback and poor growth, then production and fruit quality are reduced. Iron deficiency is usually caused by high soil pH (alkaline-above 8.0) or very wet soils. The lack of oxygen that results from the excess water inhibits root function and iron absorption by plants. Iron chlorosis may be also caused by high lime content in the soil (lime induced chlorosis). The iron-starved plants have short and slender stalk. Their terminal buds remain alive but their new leaves show chlorosis of tissues in between the veins, which themselves remains green. In vegetable crops the new leaves develop light yellow colour in between the veins. Later on the entire leaf becomes yellow. In legumes the leaves turn yellow with the veins remaining green, and on leaves spots of dead tissues appear, particularly at the margins.

3.8 Zinc

In fruit trees zinc deficiency causes small, terminal leaves with yellow mottling between the large leaf veins. Leaves near the growing tips are small (little leaf) and bunched together (rossetting). In the spring, shoots take on a yellowish cast and "mottle leaf" is quite evident. Dieback of twigs may occur in severe zinc deficiency. Zinc-deficient trees may have poor and/or delayed bloom and small fruit of low quality. Foliar sprays and chelated formulations are available to combat the problem. Timing depends on the commodity. An early to mid-november application is recommended for almonds, apples, apricots, cherries, pears, plums, and prunes. Leaf burn and defoliation may occur as a result of this spray, but these are not detrimental at this time of year. (The dormant zinc spray should not be made at the same time as a dormant oil spray.). Various plant species show different symptoms of zinc deficiency. On maize seedlings white bud disease is noticed. It is a type of chlorosis or fading of dark-green colour. These are small white spots of inactive or dead tissue. The leaves of opening buds have white or light yellow colour. Hence, the zinc deficiency disease is called 'white bud' disease. Potato plants without zinc form grayish-brown to bronze-coloured irregular spots, usually appearing in the middle of the leaves. The affected tissue sinks and finally dies. Extreme deficiency of zinc manifests in chlorotic conditions and in darker coloured veins of leaves. It is difficult to distinguish these symptoms under field conditions. In vegetable crops the new leaves have mottled appearance with yellow colour. In acute cases the necrotic or dead areas are found on new leaves. Zn-deficiency in rice causes *Khaira disease of rice*.

3.9 Manganese

In fruit trees deficiency of manganese appears as inter-veinal chlorosis with a herring bone pattern affecting only basal leaves; terminal leaves are not affected. It is usually associated with high pH soils. Reducing soil pH and foliar treatments with chelated manganese will correct this problem for several seasons. In this case also the symptoms are localized to terminal buds which remain alive, but the bud leaves are chlorotic with veins light or dark-green. In cereals the leaves turn brown or transparent, this is followed by necrosis of the affected tissues. In legumes the terminal buds remain alive but leaves become light green or yellow with green veins. Later on dead tissues appear on the leaf. Symptoms of manganese deficiency are popularly known as: *grey speck* of oats, speckled yellow of sugarbeet, marsh spot of pea. In potato the terminal buds remain alive, chlorosis of newer tissue occurs and numerous small brown patches develop which in time become more extensive. In vegetables the new leaves become chlorotic while veins remain green. Although in citrus the leaves have normal shape and size their veins remain green while the tissue in between becomes light green to grey in colour.

3.10 Copper

Male flower sterility, delayed flowering and senescence are the most important effects of Cu-deficiency. Chlorosis of the younger shoot tissues, white tips, reclamation disease, necrosis,leaf distoration, and die back are the characteristics

Cu-deficiency.The terminal buds remain alive but wilting or chlorosis of bud leaves takes place with or without spots of dead tissues. The veins of these leave remain light or dark-green. Deficiency in potato is recognized by the wilting of young leaves and loss of turgor of terminal buds which drop when flower buds are developing. There is no pronounced chlorosis but drying of leaf tips occurs in advanced stages. In vegetables the growth is retarded and leaves lack turgidity. They exhibit chlorosis as if they are bleached. In legumes the young leaves wilt with or without chlorosis. In extreme deficiency there may occur excessive leaf shedding. In citrus the large leaves are frequently malformed, and have a fine network of green veins on a light green background. The fruits have gummy excrescences.

3.11 Boron

Peach and nectarine, two of the most sensitive crops to boron toxicity, exhibit the following symptoms: small necrotic spots on the under side of the midrib, cankers along the midrib, on petioles, and on young twigs, leaf yellowing, defoliation, twig dieback and gumming in severe cases, and distorted fruits.Correction is achieved by leaching boron out of the root zone. If you suspect boron toxicity, check your irrigation water; it should not be above 2-5ppm. Additional N fertilizer as calcium nitrate may help to alleviate boron toxicity. The deficiency symptoms of this nutrient are usually localized on new or bud leaves of the plant. In corn the younger leaves are dwarfed. Their tissues are white and the growing tips dead. Under field conditions the plants have weaker ear-shanks and stalks. Their leaves are yellowish in colour. In the potato fields boron-deficiency symptoms occur in the tubers rather than on the veins. The tubers on boiling show much sloughing, are fairly saggy and have a flat flavour. In sand culture devoid of boron, the plants are short and bushy. The growing points are soon killed and the growth of lateral buds is stimulated. The leaves thicken and margins roll upwards. The leaf points and margin of older leaves die prematurely. The tubers, besides being small in size, have a ruptured surface. In vegetables the growing tissues of stems and roots are involved. The new bud leaves and petioles have light colour, are brittle and are often deformed in shape. Rosetting due to short internodes is pronounced at the shoot terminals. The legumes also have rosetting at the terminal buds. The buds appear as white or light brown dead tissue. The plants have little flowering. In citrus the deficiency symptoms are localized to new growth. New leaves have water-soaked flecks, which become translucent. The fruits have hard, fumy lumps in the rind. In deciduous trees symptoms appear on terminal tissues of twigs. The young leaves have chlorotic appearance and are wrinkled. Due to severe deficiency the twigs and spurs show symptoms of dieback. In vegetable crops the new leaves have mottled appearance with yellow colour. In acute cases the necrotic or dead areas are found on new leaves. Many deficiency diseases, such as *internal cork of apples, heart rot of sugarbeet, top rot of tobacco* and *cracked stem of celery*, are caused by its deficiency.

3.12 Molybdenum

Molybdenum deficiencies resemble the N-deficiencies. In *Brrassica* species, symptoms appear on 3 to 4 week old plants. The Mo-deficient cauliflower plants exhibit chlorotic mottling and cupping of the middle leaves. The deficiency is markedly evident in legumes, particularly in the subterranean clover. Molybdenum-starved plants have yellowish to pale-green colour. The deformity of *whiptail* produced in cauliflower is due to the deficiency of molybdenum.

3.13 Chlorine

Deficiency symptoms of chlorine include wilting of leaves, curling of leaflets and chlorosis are similar to that of Mn-deficiency.

Table 3.1: Prominent nutrient deficiency symptoms in plants

Nutrient	Colour change in lower leaves
N	Plant light green, older leaves yellow
P	Plants dark green with purple cast, leaves and plants small
K	Yellowing and scorching along the margin of older leaves
Mg	Older leaves have yellow discolouration between veins-finallyreddish purple from edge inward
Zn	Pronounced interveinal chlorosis and bronzing of leaves
Nutrient	**Colour change in lower leaves** (Terminal bud dies)
Ca	Delay in emergence of primary leaves, terminal buds deteriorate
B	Leaves near growing point turn yellow, growth buds appear as whiteor light brown, with dead tissue.
Nutrient	**Colour change in upper leaves** (Terminal bud remains alive)
S	Leaves including veins turn pale green to yellow, first appearance inyoung leaves.
Fe	Leaves yellow to almost white, interveinal chlorosis at leaf tip
Mn	Leaves yellowish-gray or reddish, gray with green veins
Cu	Young leaves uniformly pale yellow. May wilt or wither without chlorosis
Mo	Wilting of upper leaves, then chlorosis
Cl	Young leaves wilt and die along margin

Chapter-4

Soil Testing

Soil testing is well recognized as a sound scientific tool to assess inherent power of soil to supply plant nutrients. A soil test is a chemical extraction of a soil sample to estimate nutrient availability. Soil test extracts part of the total nutrient content that is related to (but not equal to) the quantity of plant available nutrient. Thus, a soil test represents only an index of the nutrient availability. Compared with plant analysis, soil test determines relative nutrient status before planting. It also includes testing of soils for other properties like pH, electrical conductivity, cation exchnage capacity, texture, structure, water holding capacity and parameters for amelioration of chemically deteriorated soils for recommending soil amendments, such as, gypsum for alkali soils and lime for acid soils. *One of the objectives of soil testing is to sort out the nutrient deficient areas from non-deficient ones.* This information is important for determining whether the soils could supply adequate nutrients for optimum crop production or not. The concept of balanced nutrition of crops also guides the use of plant nutrients in a definite proportion as required by the crops which is possible only if one knows the available nutrient status of his soils. Soil testing helps in understanding the inherent fertility status of the soils. Further, various factor other than poor soil fertility may also be responsible for poor crop production but soil fertility status assumes a greater importance. Each fertilizer recommendation based on a soil analysis should take into account the soil test value obtained by the accurate soil analysis, the research work conducted on a crop response to fertilizer application in a particular area and the practices and level of management of the concerned farmer. The soil test aimed at soil fertility evaluation with resulting fertilizer recommendation is, therefore, the actual connecting link between the agronomic research and its practical application to the farmers' field.

4.1 Objectives of Soil Testing

I. To provide an index of nutrient availability in soil. Soil test or extractant is designed to extract a portion of the nutrient from the same pool (i.e. solution, exchange, organic, or mineral) used by the plant.

II. To provide the probability of obtaining a profitable response to fertilizer. Although a response to applied nutrient will not always be obtained on low-testing soils because of other limiting factors, the probability of a response is greater than on high –testing soils.

III. To provide a basis for development of fertilizer recommendations which can be done by careful laboratory, greenhouse and field studies.

4.2 Soil Nutrient as an Index of Soil Fertility

Generally, the soil testing laboratories use organic carbon as an index of available N, Olsen's and Bray's method for available P and neutral normal ammonium acetate method for available K. In semi-arid tropics, nitrate N is also used as an index of available N in soil. Available nutrient status in the soils is generally classified as low, medium and high which are generally followed at the national level and are as follows

Table 4.1: Available soil nutrient status ratings

S. No.	Soil Nutrients	Soil Fertility Ratings		
		Low	Medium	High
1	Organic Carbon as a measure of available Nitrogen (%)	<0.5	0.5-0.75	>0.75
2	Available N as per alkaline permanganate method (kg ha^{-1})	<280	280-560	>560
3	Available P by Olsen's method (kg ha^{-1})	<10	10-24.6	>24.6
4	Available K by Neutral N^{-} ammonium acetate method	<108	108-280	>280

4.3 STCR (Soil Test Crop Response)

Soil fertility evaluation helps the farmers to use fertilizers according to the needs of the crop. Fertilizer recommendations are usually given for different crops by taking into consideration only the available nutrient status of the soil prior to raising the crop. The soil is categorized as low, medium and high fertility classes. These are generalized recommendations and do not take into account the large-scale variations from field to field. This lacuna led to the development of prescription based fertilizer recommendations for a given soil-crop-fertilizer situation. This approach takes into account the soil fertility status as well as the crop needs thus providing balanced nutrition of crops. The Soil Test Crop Response (projects) of ICAR including one for micro nutrients which were initiated in 1967 and many State Agricultural Universities (SAU's) are engaging in refining the limits and categories of soil classification. It is important to note that over the decade, only three levels of available N, P and K as determined with testing method as indicated above, continued to most operative. In many situations, the testing method and limit fixed for available K is found not to be satisfactory while nitrogen content continues to be represented by organic carbon which at times has no direct relation with soil available nitrogen. The broad classification of soil nutrients status is too general and may be only indicative for national level appreciation of soil fertility status and not for the benefit of the individual farmer. This classification needs refinement.

Micro-nutrient deficiencies also started becoming critical, beginning with the intensification of agriculture and using mostly high analysis chemical fertilizers. Thus micronutrient testing facilities were also required to be created in the soil testing laboratories.

4.3.1 Soil Testing for Micro and Secondary Nutrients

To ensure that deficient micro and secondary nutrients are supplied to the crops, a systematic delineation has been initiated by All India Coordinated Research Projects of micro-nutrients in Soil and Plants since renamed as Micro, secondary and pollutant elements in soils & plants. Testing method for assessing available micronutrients has been standardized and is being used in soil testing laboratories. The amount of the nutrients that can remove yearly by the normal crop yields vary from element to element and crop to crop. The form of the nutrients and their mobility in the plants are given in the table 4.2 and the critical limits in soil, plant and soil tests are given in the table 4.3

Table 4.2: The form of the nutrients and their Relative mobility in the plants

Element	Form Absorbed	Mobility in Plants
Sulphur	SO_4^{2-}, SO_2	less immobile
Calcium	Ca^{2+}	immobile
Magnesium	Mg^{2+}	Mobile
Boron	H_3BO_3	immobile
Copper	Cu^{2+}	less mobile
Iron	Fe^{2+}	less immobile
Manganese	Mn^{2+}	less immobile
Molybdenum	MoO_4^{2-}	less mobile
Zinc	Zn^{2+}	moderately mobile
Chlorine	Cl^-	less mobile

Plant can absorb SO_2 directly from the atmosphere.

Table 4.3: Soil test methods with critical limits in soil and plant.

Nutrients	Soil Test Method	Critical level in soil	Critical level in plant
Sulphur	Hot water, $CaCl_2$	8-30 ppm	<0.15-0.2%
Calcium	Ammonium Acetate	<25% of CEC or <1.5meq Ca/100g	<0.2 %
Magnesium	Ammonium Acetate	<4 % of CEC or <1 meq Mg/100g	<0.1-0.2%
Boron	Hot water	0.5 ppm	<20 ppm
Copper	DTPA or Ammonium Acetate	0.2 ppm	<4 ppm (3-10)
Iron	DTPA Ammonium Acetate	2.5 – 4.5 ppm	<50 ppm(25– 80)
Manganese	DTPA	2 ppm	<20ppm (10-30)
Molybdenum	acid ammonium oxalate	0.2 ppm	<0.1 ppm
Zinc	DTPA	0.6(0.4-1.2) ppm	<15-20 ppm

4.4 Soil Fertility Evaluation Techniques

Quantifying the nutrient requirements by soil and plant analysis depends upon the careful sampling and analytical methods calibrated for the representative crops and soils in a specific region. Knowing the relationship between soil test result and crop nutrient response is essential for providing accurate recommendation. Several techniques are commonly employed to assess the nutrient status of a soil.

4.4.1. Visual nutrient deficiency symptoms

4.4.2. Biological tests

4.4.3. Plant analysis

4.4.4 Soil testing

4.4.1 Visual Nutrient Deficiency Symptoms

Careful visual observation of the growing plants can help to identify a specific nutrient stress. A nutrient-deficient plant will exhibit characteristic symptom because normal plant processes are inhibited. Visual evaluation of nutrient stress should be used to support or direct other diagnostic techniques (i.e., soil and plant analysis). It may be pointed out that deficiency symptoms in many cases are not always clearly defined and in cases, the symptoms can be common to other causes or may be masked by other nutrients. Some typical examples are given below:

1. N deficiency can be confused with S deficiency.
2. Calcium deficiency can be confused with B deficiency
3. Fe deficiency can be confused with Mn deficiency
4. Leaf stripe disease of oat can be confused with Mn deficiency
5. Effect of virus, little leaf etc. can be confused with zinc deficiency, B deficiency.
6. Brown streak disease of rice can be confused with Zn deficiency.

4.4.2 Biological Tests

a) ***Field tests:*** The field plot technique essentially measures the crop response to nutrients. In this, specific treatments are selected, randomly assigned to an area of land, which is representative of the conditions. Several replications are used to obtain more reliable results and to account for variations in soil. Field experiments are essential in establishing the equation used to provide fertilizer recommendations that will optimize crop yield, maximize profitability and minimize the environmental impact of nutrient use. When large number of tests are conducted on soils that are well characterized, recommendations can be extrapolated to other soils with similar characteristics. Field tests are expensive and time-consuming, however, they are valuable tools and are widely used by scientists. They are used in conjunction with laboratory and greenhouse studies in calibration of soil and plant tests.

b) ***Greenhouse tests:*** The greenhouse techniques utilize small quantities of soil to quantify the nutrient-supplying power of a soil. Generally, soils are collected to represent a wide range of soil chemical and physical properties that contribute to the variation in availability for a specific nutrient. Selected treatments are applied to the soils and a crop is planted that is sensitive to the specific nutrient being evaluated. Crop response to the treatments can then be determined by measuring total plant yield and nutrient content.

c) ***Laboratory tests***: Neubauer seedling method is based on the uptake of nutrients by a large number of plants grown on a small amount of soil. The roots thoroughly penetrate the soil, exhausting the available nutrient supply within a short time. The total nutrients removed are quantified and critical limits are established to give the minimum values of macro and micro-nutrients available for satisfactory yields of various crops.

d) ***Microbiological methods:*** In the absence of nutrients, certain micro-organisms exhibit behaviour similar to that of higher plants. For example, growth of *Azotobacter* or *Aspergillus niger* reflects nutrient deficiency in the soil. The soil is rated from very deficient to not deficient in the respective elements, depending on the amount of colony growth. In comparison with methods that utilize higher plants, microbiological methods are rapid, simple and require little space. These quick laboratory tests are not in common use in India.

4.4.3 Plant Analysis

This involves two approaches, one is analysis of plant in the laboratory and the other is tissue test fresh in the field. The basis on plant analysis is that amount of a given nutrient in a plant is directly related to availability of the nutrient in the soil. Plant analysis is used to:

1. Confirm a diagnosis made from visible symptoms.
2. Identify hidden hunger where no symptoms appear.
3. Locate soil areas where deficiencies of one or more nutrients occur.
4. Determine whether applied nutrients are taken by plant.
5. Indicating interactions or antagonistic effect among nutrient elements.
6. Study the internal functioning of nutrients in plants.
7. Suggest additional tests or studies in identifying a crop production problem.

Plant analysis is used more for fruit and vegetable crops. Because of the perennial nature of the fruit crops and their extensive root systems, plant analysis is especially suitable for determining their nutrient status. As more is known about plant nutrition and nutrient requirements throughout the season, and as application of nutrients through irrigation systems is possible, plant analysis assumes greater importance. Also, to produce high yields, it is helpful in monitoring the plant through its growing season. It is becoming more useful for field crops also the critical nutrient concentration (CNC) is used in interpreting plant analysis results and diagnosing nutritional problems. CNC is the level of a nutrient below which crop yield or quality is unsatisfactory. In India, a great deal of information has been generated on critical nutrient level in crops especially for micro-nutrients and sulphur. The demand for this service should increase in India in future as research emphasizes opportunities to manage nutrient availability during the growing season an important phase of plant analysis is sample collection. Plant composition varies with age, the portion of the plant sampled, the condition of the plant, the variety, the weather and other factors. Therefore, it is necessary to follow proper sampling instructions. For plant analysis, a typical plant part is selected which indicates the nutrient status with crop to determine whether the crop needs fertilization. The optimum values are pre-standardized so as to make recommendation after sample analysis. Growth stage is important in plant analysis because nutrient status and demand varies during the season. Nutrient concentrations in the vegetative parts usually decrease with maturity. Misinterpretations of the plant analysis results are common if sampling time is not identified correctly.

Table 4.4: Plant sampling guidelines for selected agricultural and horticultural crops

Crop	Sampling Time	Plant Parts Sampled	Sample
Field crops			
Mustard	Before seed set	Newest mature leaf	20-30
Sunflower	Before heading	Newest mature leaf	20-30
Maize	Seedling stage	All above ground material	25-30
Alfalfa	Early bloom	Top 6 in. or upper 1/3rd of plant	20-30
Small Grains (barley, oats, wheat, rye, rice)	Seedling stage Before heading	All above ground material Uppermost leaf blade	25-40
Sorghum	Before or at heading	Leaf from top of plant	20-30
Vegetables			
Beans	Seedling stage	All-above ground portions	20-30
Beets	Mature	Young mature leaf	20-30
Cabbage and cauliflower	Before heading	Newest mature leaf, centre of whorl	10-20
Celery	Midseason	Outer petiole of newest mature leaf	12-20
Cucumber	Before fruit set	Newest mature leaf	12-20
Eggplant/Brinjal	Early fruiting	Young mature leaf	12-20
Garlic	Bulbing	Young mature leaf	15-25
Lettuce, spinach	Midseason	Newest mature leaf	15-25
Melons	Before fruit set	Newest mature leaf	15-25
Peas	Before or at bloom	Leaves from node from top	40-60
Peppers	Mid-season	Recently mature leaf	25-50
Potato	Before or at bloom	Leaf from growing tip	25-30
Pumpkin/Squash	Early fruiting	Young mature leaf	15-25
Radish	Mid-growth to harvest	Young mature leaf	40-50
Root crop (carrot, beet, onion)	Before root or bulb enlargement	Newest mature leaf	20-30
Sweet potato	Mid-season, before root enlargement	Leaf from tip centre	20-30
Tomato	Early mid bloom	Leaf from growing tip	15-25
Ornamentals			
Carnation	Newly planted	4th-lished5th pair from base of plant	20-30
	Established	5th -6th pair from base of plant	20-30

Contd...

Table 4.4: Contd...

Crop	Sampling Time	Plant Parts Sampled	Sample
Chrysanthemum	Before or at bloom	Top leaves on flowering stem	20-30
Poinsettia	Before or at bloom	Newest mature leaf	15-25
Rose	At bloom	Newest mature compound leaf on flowering stem	25-30
Trees and Shrubs	Current year growth	Newest mature leaf	30-70
Turf	Active growth	Leaf blade-avoid soil contamination	2 cups
Fruits and Nut crops			
Apple, pear, almond , apricot, cherry,plum and prune	Mid-season	Middle leaves from current year growth-non fruiting, non-expanding spurs	50-100
Grapes	At bloom	Petiole or leaves adjacent to basal cluster at bloom	-
Peach, nectarine	Mid-season	Midshoot leaflets/leaves	25-100
Walnut	6-8 weeks after bloom	Terminal leaflets/leaves from non-fruiting shoots	25-40
Strawberries	Mid-season	Newest mature leaf	25-40
Raspberries	Mid-season	Newest mature leaf, laterals from primocanes	30-50
Pecan	Mid-season	Mid-shoot leaflets/ leaves	25-60

4.4.4 Soil Testing

The importance of soil testing can be appreciated in ensuring balanced fertilization for a profitable and sustainable crop production. In view of the importance of balanced fertilization in increasing crop production, the following recommendations were made in the FAI's Annual (2009) International Seminar on Fertilizer Policy for Sustainable Agriculture which is widely attended by the State Governments, fertilizer industries and experts.

Recommendation No.9: It relates to soil health and it is stated that "Soil specific nutrient management, which is considered as Best Fertilizer Management Practices (BFMPs), needs to be promoted to improve soil health and crop productivity."

Recommendation No.11: "There is an imminent threat to India's food security arising out of imbalanced use of fertilizers, absence of component-wise locations / crop specific fertilizer recommendations and limited use of organic fertilizers, widespread deficiencies of secondary and micro-nutrients, diminishing effect of crop response to fertilizers, limited irrigation resources, stagnation of area under cultivation and stagnating food grain productivity."

Recommendation No.12: It is stated that "Inadequate and unreliable soil testing facilities, poor awareness of farmers about balanced plant nutrition and lack of appropriate policy are the major constraints in adoption of "Fertilizer Best Management Practices." The wording of recommendation (No.12) may be considered too critical but the fact is realized that the soil testing service has not made the desired impact and farmers have not yet been able to adopt it in large numbers. While at the same time, the importance of the soil test based balanced fertilizer use is being advocated by the Government, Fertilizer Industries and others concerned. The quality of the soil testing programme, therefore, needs to be improved in all its aspects.

4.5 Balanced Fertilization

Balanced fertilization does not mean a certain definite proportion of Nitrogen, Phosphorus and Potash (or other nutrients) to be added in the form of fertilizers and organic manures but it has to take into account the availability of nutrients already present in the soil, crop requirement and other factors. Generally, referred 4:2:1 N:P:K use as a desired ratio does not substitute the need and importance of actually working out of the nutrient deficiency in the soil and addition of required nutrients through fertilizers and manures to meet the crop need. This ratio actually relates to the general fertilizer use recommendation started for the two major cereal crops, i.e. rice and wheat as 120: 60: 30 kg NPK per hectare (4: 2: 1). Till the beginning of green revolution era, farmers have been traditionally using only nitrogen as the first nutrient being almost universally deficient in Indian soils which also showed conspicuous response to its application. Thus, an emphasis on a particular ratio (4 : 2 : 1) has helped in increasing the use of P&K which was necessary to be used over the application of N when high yielding varieties were introduced. Balanced fertilization should also take into account the crop removal of nutrient, crop species to be sown, farmers' investment ability, soil moisture regimes, weed control, soil salinity, alkalinity, physical environment, microbiological conditions of the soils which determine the status of available nutrients in soil and cropping sequence etc. The balanced fertilizer should aim at:

I. Increasing crop yield and quality.

II. Increasing farm income

III. Correction of inherent soil nutrient deficiency.

IV. Maintaining or improving soil fertility

V. Avoiding damage to the soil ecosystem and environment

VI. Restoring – fertility and productivity of the land that may have been degraded by exploitive activities in the past.

4.6 Soil Testing Procedures

Procedurally, the soil testing programme can be divided into the following components:

4.6.1 Error, Precision, Accuracy and Detection Limit

An enormous variety of definitions relating to detection limits and quantitation limits are used in the literature and by government agencies. Unfortunately, universally accepted procedures for calculating these limits do not exist. This can be frustrating and confusing for both regulators and the regulated community. Following are some Numerical Criteria for selecting an analytical method and their definitions to resolve these issues:

4.6.1.1 Accuracy

It is a combination of the bias and precision of an analytical procedure, which reflects the closeness of a measured value to a true value. (Standard Methods, 18th edition). For the purposes of laboratory certification, accuracy means the closeness of a measured value to its generally accepted value or its value based upon an accepted reference standard. (NR 149.03(2)).Bias provides a measure of systematic or determinative error in an analytical method. Bias is determined by assessing the percent recovery of spiked samples. Historically, the term **accuracy** has been used interchangeably with bias, although many sources make a distinction between the two (Standard Methods, 18th edition).

4.6.1.2 Enforcement Standard (ES)

It means a numerical value expressing the maximum concentration of a substance in ground water which is adopted under s.160.07, Stats, and s. NR 140.10 or s. 160.09, Stats. and s. NR 140.12. These standards are toxicologically derived to protect human health. Analytical values above the ES trigger the procedure prescribed in s. NR 140.26. (s.NR 140.05(7)) False Positive, or Type I (alpha) error, means concluding that a substance is present when it truly is not. False Negative, or Type II (beta) error, means concluding that a substance is not present when it truly is.

4.6.1.3 Instrument Detection Limit (IDL)

It is the concentration equivalent to a signal, due to the analyte of interest, which is the smallest signal that can be distinguished from background noise by a particular instrument. The IDL should always be below the method detection limit, and is not used for compliance data reporting, but may be used for statistical data analysis and comparing the attributes of different instruments. The IDL is similar to the "critical level" and "criterion of detection" as defined in the literature. (Standard Methods, 18th edition)

4.6.1.4 Limit of Detection (LOD)

LOD or detection limit, is the lowest concentration level that can be determined to be statistically different from a blank (99% confidence). The LOD is typically determined to be in the region where the signal to noise ratio is greater than 5. Limits of detection are matrix, method, and analyte specific. (ss. NR 140.05(12) & 149.03(15))

Note: For the purposes of laboratory certification, the LOD is approximately equal to the MDL for those tests which the MDL can be calculated.

4.6.1.5 Limit of Quantitation (LOQ)

Lower limit of quantitation (LOQ) is the level above which quantitative results may be obtained with a specified degree of confidence. The LOQ is Analytical Detection Limit Guidance mathematically defined as equal to 10 times the standard deviation of the results for a series of replicates used to determine a justifiable limit of detection. Limits of quantitation are matrix, method, and analyte specific. (ss. NR 140.05(13) & 49.03(16))

4.6.1.6 Linear Calibration Range (LCR)

Range of Linearity (LCR) is the region of a calibration curve within which a plot of the concentration of an analyte versus the response of that particular analyte remains linear and the correlation coefficient of the line is approximately 1 (0.995 for most analytes). The plot may be normal-normal, log-normal, or log-log where allowed by the analytical method. At the upper and lower bounds of this region (upper and lower limits of quantitation), the response of the analyte's signal versus concentration deviates from the line. The concentration limit of quantification (LOQ) to the limit of linearity (LOL) is also called as *Dynamic Range.*

4.6.1.7 Maximum Contaminant Level (MCL)

It is a numerical value expressing the maximum permissible level of a contaminant in water which is delivered to any user of a public water system. Maximum contaminant levels are listed in s. NR 809.09, Wis. Adm. Code. (NR 809.04(34)).

4.6.1.8 Method Detection Limit (MDL)

It is the minimum concentration of a substance that can be measured and reported with 99% confidence that the analyte concentration is greater than zero, and is determined from analysis of a sample in a given matrix containing the analyte.

4.6.1.9 Practical Quantitation Limit (PQL)

It is a quantitation limit that represents a practical and routinely achievable quantitation limit with a high degree of certainty (>99.9% confidence) in the results.

4.6.1.10 Precision

It is a measure of the random error associated with a series of repeated measurements of the same parameter within a sample. Precision describes the closeness with which multiple analyses of a given sample agree with each other, and is sometimes referred to as reproducibility. Precision is determined by the absolute standard deviation, relative standard deviation, variance, coefficient of variation, relative percent difference, or the absolute range of a series of measurements. (s. NR140.05 (16) and Standard Methods, 18th edition). Mathematically can be calculated as:

Terms	Definition
Absolute standard deviation, s	$s = \sqrt{\frac{\Sigma_t^N = (x_i - \bar{x})^2}{N-1}}$
Relative standard deviation (RSD)	$RSD = \frac{S}{x}$
Standard error of the mean, s_m	$S_m = \frac{s}{\sqrt{N}}$
Coefficient of variation (CV)	$CV = \frac{s}{x} \times 100\%$
Variance	s^2

$$Where, \bar{x} = \sqrt{\frac{\sum_{i=1}^{N} X_i}{N}}$$

4.6.1.11 Preventive Action Limit (PAL)

It is a numerical value expressing the maximum concentration of a substance in groundwater which is adopted under s. 160.15, Stats., and s. NR 140.10, 140.12 or140.20. Reported values above the PAL trigger the procedure prescribed in s. NR 140.22. The PAL is typically set at 1/10th of the enforcement standard if the substance is carcinogenic, mutagenic, and teratogenic or has a synergistic effect. The PAL is 20% of the enforcement standard for other substances of public health concern. Reporting Limit is an arbitrary number below which data is not reported. The reporting limit mayor may not be statistically determined, or may be an estimate that is based upon the experience and judgment of the analyst. Analytical results below the reporting limit are expressed as "less than" the reporting limit. Reporting limits are not acceptable substitutes for detection limits unless specifically approved by the Department for a particular test.

Sample Matrix, or Matrix, defines the general physical-chemical makeup of a particular sample. Although the actual matrix of a sample varies from discharge to discharge and from location to location, general classes of matrices include; reagent (clean) water, wastewater, public drinking water, waste, surface water, groundwater, sediments and soils. (NR 149.03(28)).

4.6.1.12 Sample Standard Deviation, or Standard Deviation (s),

Is a measure of the degree of agreement, or precision, among replicate analyses of a sample. (Standard Methods, 18th edition) In this document, standard deviation implies sample standard deviation (n-1 degrees of freedom). The population standard deviation (n degrees of freedom) should only be used when dealing with a true approximation of a population (e.g. greater than 25 data points). The standard deviation is defined as:

i) Sensitivity means the ability of a method or instrument to detect an analyte at a specified concentration. (NR 149.03(28m)).

ii) Signal to Noise Ratio (S/N) is a dimensionless measure of the relative strength of an analytical signal (S) to the average strength of the background instrumental noise (N) for a particular sample and is closely related to the detection level. The ratio is useful for determining the effect of the noise on the relative error of a measurement. The S/N ratio can be measured a variety of ways, but one convenient way to approximate the S/N ratio is too divide the arithmetic mean (average) of a series of replicates by the standard.

4.7 Quality Control of Analytical Procedures

4.7.1 Independent Standards

The ultimate aim of the quality control measures is to ensure the production of analytical data with a minimum of error and with consistency. Once, an appropriate method is selected, its execution has to be done with utmost care. To check and verify the accuracy of analysis, independent standards are used in the system. The extent of deviation of analytical value on a standard sample indicates the accuracy of the analysis. Independent standard can be prepared in the laboratory from pure chemicals. When new standard is prepared, the remainders of the old ones always have to be measured as a mutual check. If the results are not within the acceptable levels of accuracy, the process of calibration, preparation of standard curve and the preparation of reagents may be repeated till acceptable results are obtained on the standard sample. After assuring this, analysis on unknown sample has to be started. Apart from independent standard, certified reference samples can also be used as 'standard'. Such samples are obtained from other selected laboratories where the analysis on a prepared standard is carried out by more than one laboratory and such samples along with the accompanied analytical values are used as a check to ensure the accuracy of analysis.

4.7.2 Use of Blank

A blank determination is an analysis without the analyte or attribute or in other words, an analysis without a sample by going through all steps of the procedure with the reagents only. Use of the blank accounts for any contamination in the chemicals used in actual analysis. The 'estimate' of the blank is subtracted from the estimates of the samples. The use of 'sequence control' samples is made in long batches in automated analysis. Generally two samples, one with a low content of analyte and another with very high content of known analyte (but the contents falling within the working range of the method) are used as standards to monitor the accuracy of analysis.

4.7.3 Blind Sample

A sample with known content of analyte, is inserted by the head of the laboratory in batches and times unknown to the analyst. Various types of sample material may serve as blind samples such as control samples or sufficiently large leftover of test samples (analysed several times). It is essential that analyst is aware of the possible presence of a blind sample but is not able to recognize the material as such.

4.7.4 Validation of Procedures of Analysis

Validation is the process of determining the performance characteristics of a method / procedure. It is a pre-requisite for judgment of the suitability of produced analytical data for the intended use. This implies that a method may be valid in one situation and invalid in another. If a method is very precise and accurate but expensive for adoption, it may be used only when the data with that order of precision are needed. The data may be inadequate, if the method is less accurate than required. Two types of validation are followed.

4.7.5 Validation of Own Procedure

In-house validation of method or procedure by individual user laboratory is a common practice. Many laboratories use their own version of even well established method for reasons of efficiency, cost and convenience. A change in liquid solid ratio in extraction procedures for available soil nutrients and shaking time etc. result in changed value, hence need validation. Such changes are often introduced to consider local conditions, cost of analysis, required accuracy and efficiency. Validation of such changes is the part of quality control in the laboratory. It is also a kind of research project, hence all types of the laboratories may not be in a position to modify the standard method. They should follow the given method as accepted and practiced by most other laboratories. Apart from validation of methods, a system of internal quality control is required to be followed by the laboratories to ensure that they are capable of producing reliable analytical data with minimum of error. This requires continuous monitoring of the operation and systematic day to day checking of the produced data to decide whether these are reliable enough to be released.

Following steps need to be taken for internal quality control:

- Use a blank and a control (standard) sample of known composition along with the samples under analysis.
- Round off the analytical values to the 2nd decimal place. The value of 3rd decimal place may be omitted if less than 5. If it is more than 5, the value of second decimal may be raised by 1. Since the quality control systems rely heavily on control samples, the sample preparation may be done with great care to ensure that the:
  - Sample is homogenous.
  - Sample material is stable.
  - Sample has uniform and correct particle size as sieved through a standard sieve.
  - Relevant information such as properties of the sample and the concentration of the analyte are available.

The samples under analysis may also be processed / prepared in such a way that it has similar particle size and homogeneity as that of the standard (control) sample. As and when an error is noticed in the analysis through internal check, corrective measures should be taken. The error can be due to calculation or typing. If not, it requires thorough check on sample identification, standards, chemicals, pipettes, dispensers, glassware, calibration procedure and equipment. Standard may be old or wrongly prepared. Pipette may indicate wrong volume, glassware may not be properly cleaned and the equipment may be defective or the sample intake tube may be clogged in case of flame photometer or Atomic Absorption Spectrophotometer. Source of error may be detected and samples be analyzed again.

4.7.6 Validation of the Standard Procedure

This refers to the validation of new or existing method and procedures intended to be used in many laboratories including procedures accepted by national system or ISO. This involves an inter-laboratory programme of testing the method by a member of selected renowned laboratories according to a protocol issued to all participants. Validation is not only relevant when non-standard procedures are used but just as well when validated standard procedures are used and even more so when variants of standard procedures are introduced. The results of validation tests should be recorded in a validation report from which the suitability of a method for a certain purpose can be deduced.

4.7.7 Inter-laboratory Sample and Data Exchange Programme

If an error is suspected in the procedure and uncertainty cannot readily be solved, it is not uncommon to have the sample analysed in another laboratory of the same system/organization. The results of the other laboratory may or may not be biased, hence doubt may persist. The sample check by another accredited laboratory may be necessary and useful to resolve the problem. An accredited laboratory should participate at least in one inter-laboratory exchange programme. Such programmes do exist locally, regionally, nationally and internationally.

The laboratory exchange programme exists for method performance studies and laboratory performance studies. In such exchange programme, some laboratories or the organizations have devised the system where periodically samples of known composition are sent to the participating laboratory without disclosing the results. The participating laboratory will analyse the sample by a given method and find out the results. It provides a possibility for assessing the accuracy of the method being used by a laboratory, and also about the adoption of the method suggested by the lead laboratory. For quality check, each laboratory will benefit if it becomes part of some sample/method check and evaluation programme. The system of self-check within the laboratory also has to be regularly followed.

4.7.8 Preparation of Reagent Solutions and their Standardization

Chemical reagents are manufactured and marketed in different grades of purity. In general the purest reagents are marketed as Analytical Reagent or "AR" grade. Further, the markings may be "LR" meaning laboratory reagent or "CP", meaning chemically pure. The strength of chemicals is expressed as normality or molarity. It is, therefore, useful to have some information about the strength of important acids and alkali most commonly used in the chemical laboratories. Acids and alkali are basic chemicals required in a laboratory.

4.8 Molarity

One molar (M) solution contains one mole or one molecular weight in grams of a substance in each litre of the solution. Molar method of expressing concentration is useful due to the fact that the equal volumes of equimolar solutions contain equal number of molecules.

4.9 Normality

The normality of a solution is the number of gram equivalents of the solute per litre of the solution. It is usually designated by letter (N). Semi-normal, penti-normal, desinormal, centi-normal and milli-normal solutions are often required, these are written (shortly) as 0.5N, 0.2N, 0.1N, 0.01N and 0.001N, respectively. However, molar expression is preferred because 'odd' normalities such as 0.121N are clumsily represented in fractional form.

Table 4.5 Strength of commonly used acids and alkali

S. No.	Reagent/Chemical	Normality	Molarity	Formula by weight	% by weight (approx.)	Specific gravity (approx.)	ml requiredfor 1N/litre solution(aprox.)	ml required for 1M/liter Solution (aprox.)
1.	Nitric acid	16	16	63	70	1.42	63.7	63.7
2.	Sulphuric acid	35	17.5	98	98	1.84	28	56
3.	Hydrochloric acid	11.6	11.6	36.5	37	1.19	82.6	82.6
4.	Phosphoric acid	45	15	98	85	1.71	22.7	68.1
5.	Perchloric acid	10.5	10.5	100.5	65	1.6	108.7	108.7
6.	Ammonium Hydroxide	15	15	35	28	0.90	67.6	67.6

The definition of normal solution utilizes the term 'equivalent weight'. This quantity varies with the type of reaction, and hence it is difficult to give a clear definition of equivalent weight which will cover all reactions. It often happens that the same compound possess different equivalent weights in different chemical reactions. A situation may arise where a solution has normal concentration when employed for one purpose and a different normality when used in another chemical reaction. Hence, the system of molarity is preferred.

4.10 Equivalent Weight

The equivalent weight of a substance is the weight in grams which in its reaction corresponds to a gram atom of hydrogen or of hydroxyl or half a gram atom of oxygen or a gram atom of univalent ion. When one equivalent weight of a substance is dissolved in one litre, it gives I N solution. Equivalent and molecular weights of some important compounds are given in the following table 4.6

Table 4.6: Equivalent and molecular weights of some important compounds

Compound	Formulae	Mol Wt.(g)	Eq. wt (g)
Ammonium acetate	$NH_4C_2H_3O_2$	77.08	77.08
Ammonium chloride	NH_4Cl	53.49	53.49
Ammonium fluoride	NH_4F	37.04	37.04
Ammonium nitrate	NH_4NO_3	80.04	80.04
Barium chloride	$BaCl_2.2H_2O$	244.28	122.14
Boric acid	H_3BO_3	61.83	20.61
Calcium acetate	$(CH_3COO)_3Ca$	158.00	79.00
Calcium Carbonate	C_aCO_3	100.09	50.05
Calcium chloride (dehydrate)	$CaCl_2.2H_2O$	147.02	73.51
Calcium hydroxide	$Ca(OH)_2$	74.00	37.00
Calcium nitrate	$Ca(NO_3)_2$	164.00	82.00
Calcium sulphate	$CaSO_4.2H_2O$	172.17	86.08
Ferrous ammonium sulphate	$(NH_4)_2SO_4.FeSO_4.6H_2O$	392.13	392.13
Ferrous sulphate	$FeSO_4.7H_2O$	278.01	139.00
Magnesium chloride	$MgCl_2.6H_2O$	203.30	101.65
Magnesium nitrate	$Mg(NO_3)_2.6H_2O$	256.41	128.20
Potassium chloride	KCl	74.55	74.55
Potassium dichromate	$K_2Cr_2O_7$	294.19	49.04
Potassium hydroxide	KOH	158.03	158.03
Potassium permanganate	$KMnO_4$	158.03	31.60
Potassium nitrate	KNO_3	101.10	101.10
Potassium sulphate	K_2SO_4	174.27	87.13
Potassium hydrogen phthalate	COOH C_6H_6 COOK	204.22	204.22
Oxalic acid	$C_2H_2O_4.2H_2O$	126.00	63.00
Silver nitrate	$AgNO_3$	169.87	169.87

Contd...

Table 4.6: Contd...

Compound	Formulae	Mol Wt. (g)	Eq. wt. (g)
Sodium acetate(Anhydrous)	CH_3COONa	82.04	82.04
Sodium bicarbonate	$NaHCO_3$	84.01	84.01
Sodium carbonate	Na_2CO_3	106.00	53.00
Sodium chloride	NaCl	58.45	58.45
Sodium hydroxide	NaOH	40.00	40.00
Sodium nitrate	$NaNO_3$	84.99	84.99
Sodium oxalate	$Na_2C_2O_4$	134.00	67.00
Sodium sulphate	Na_2SO_4	142.04	71.02
Sodium thiosulphate	$Na_2S_2O_3.5H_2O$	248.18	248.18

4.11 Milli-equivalent Weight

Equivalent weight (Eq W) when expressed as milli-equivalent weight (meq W), means the equivalent weight in grams divided by 1000. It is commonly expressed by "meq". It is the most convenient value because it is the weight of a substance containedin or equivalent to one ml of I N solution. It is, therefore, a unit which is common to both volumes and weights, making it possible to convert the volume of a solution to its equivalent weight and the weight of a substance to its equivalent volume ofsolution.

Number of meq = Volume x Normality

4.12 Buffer Solutions

Solutions containing a weak acid and its salt or weak base and its salt (e.g.CH_3COOH + CH_3COONa) and (NH_4OH + NH_4Cl) possess the characteristic property to resist changes in pH when some acid or base is added in them. Such solutions are referred to as buffer solutions. Following are important properties of a buffer solution:

- It has a definite pH value.
- Its pH value does not alter on keeping for a long time.
- Its pH value is only slightly altered when strong base or strong acid is added.

It may be noted that because of the above property, readily prepared buffer solutions of known pH are used to check the accuracy of pH meters being used in the laboratory.

4.13 Titration

It is a process of determining the volume of a substance required to just complete the reaction with a known amount of other substance. The solution of known strength used in the titration is called titrant. The substance to be determined in the solution is called titrate. The completion of the reaction is judged with the help of appropriate indicator.

4.14 Indicator

A substance which indicates the end point on completion of the reaction is called as indicator. Most commonly used indicators in volumetric analysis are:

- internal indicator
- external indicator
- self indicator

4.14.1 Internal Indicator

The indicators like methyl red, methyl orange, phenolphthalein and diphenylamine which are added in the solution where reaction occurs, are called internal indicators. On completion of the reaction of titrant on titrate, a colour change takes place due to the presence of indicator, which also helps in knowing that the titration is complete. Typical examples of colour change due to pH change is solutions are given in Table 4.7

Table 4.7: pH Transition interval of some indicators

pH Indicator		pH transition interval		
Name	**Colour**	**pH**	**pH**	**Colour**
Cersol red	Pink	0.2	1.8	Yellow
m-Cresol purple	Red	1.2	2.8	Yellow
Thymol blue	Red	1.2	2.8	Yellow
2,4-Dinitrophenol	Colourless	2.8	4.7	Yellow
Bromochlorophenol blue	Yellow	3.0	4.7	Purple
Bromophenol blue	Yellow	3.0	4.6	Purple
Methyl orange	Red	3.1	4.4	Yellow-orange
Bromocresol green	Yellow	3.8	5.4	Blue
2,5-Dinitrophenol	Colourless	4.0	5.8	Yellow
Methyl red	Red	4.4	6.2	Yellow-orange
Chlorophenol red	Yellow	4.8	6.4	Purple
Litmus extra pure	Red	5.0	8.0	Blue
Bromophenol red	Orange-yellow	5.2	6.8	Purple
Bromocresol purple	Yellow	5.2	6.8	Purple
4-Nitrophenol	Colourless	5.4	7.5	Yellow
Bromoxylenol blue	Yellow	5.7	7.4	Blue
Bromothymol blue	Yellow	6.0	7.6	Blue
Phenol red	Yellow	6.4	8.2	Red
3-Nitrophenol	Colourless	6.6	8.6	Yellow-orange
Cresol red	Orange	7.0	8.8	Purple
1-Naphtholphthalein	Brownish	7.1	8.3	Purple
Thymol blue	Yellow	8.0	9.6	Blue
Phenolphthalein	Colourless	8.2	9.8	Red-violet
Thymolphthalein	Colourless	9.3	10.5	Blue

The internal indicators used in acid - alkali neutralization solutions are methyl orange, phenolphthalein and bromothymol blue. The indicator used in precipitation reactions like titration of neutral solution of NaCl (or chlorine ion) with silver nitrate ($AgNO_3$) solution is K_2CrO_4. On the completion of titration reaction of $AgNO_3$ with chlorine, when no more chlorine is available for reaction with silver ion to form AgCl, the chromium ions combine with Ag^{2+} ions to form sparingly soluble Ag_2CrO_4, which is brick red in colour. It indicates that chlorine has been completely titrated and end point has occurred redox indicators are also commonly used. These are substances which possess different colours in the oxidized and reduced forms. Diphenylamine has blue violet colour under oxidation state and colourless in reduced condition. Ferrocin gives blue colour under oxidation state and red colour under reduced condition.

4.14.2 External Indicator

Some indicators are used outside the titration mixture. Potassium ferricyanide is used as an external indicator in the titration of potassium dichromate and ferrous sulphate in acid medium. In this titration, few drops of indicator are placed on a white porcelain tile. A glass rod dipped in the solution being titrated is taken out and brought in contact with the drops of indicator placed on white tile. In the beginning deep blue colour is noticed which turns greenish on completion of titration.

4.14.3 Self Indicator

The titrant, after completion of the reaction leaves its own colour due to its slight excess in minute quantities. In $KMnO_4$ titration with ferrous sulphate, the addition of $KMnO_4$ starts reacting with $FeSO_4$ which is colourless. On completion of titration, slight excess presence of $KMnO_4$ gives pink colour to the solution which acts as a self-indicator and points to the completion of the titration.

4.15 Standard Solution

The solution of accurately known strength (or concentration) is called a standard solution. It contains a definite number of gram equivalent or gram mole per litre of solution. If it contains 1 gram equivalent weight of a substance per compound, it is 1N solution. If it contains 2 gram equivalent weights of the compound, it is 2N.All titrametric methods depend upon standard solutions which contain known amounts (exact) of the reagents in unit volume of the solution. A solution is prepared, having approximately the desired concentration. This solution is then standardized by titrating it with another substance which can be obtained in highly purified form. Thus, potassium permanganate solution can be standardized against sodium oxalate which can be obtained in a high degree of purity, since it is easily dried and is non-hygroscopic. Such substance, whose weight and purity is stable, is called as 'Primary Standard'. A primary standard must have the following characteristics:

- It must be obtainable in a pure form or in a state of known purity.
- It must react in one way only under the condition of titration and there must be no side reactions.

- It must be non-hygroscopic. Salt hydrates are generally not suitable as primary standards.
- Normally, it should have a large equivalent weight so as to reduce the error in weighing.
- An acid or a base should preferably be strong, that is, they should have a high dissociation constant for being used as standards.

Primary standard solution is one which can be prepared directly by weighing and with which other solutions of approximate strength can be titrated and standardized. Some Primary standards are given below:

Acids	Bases	Oxidizing agents	Reducing agents	Others
1.Potassium hydrogen phthalate	1. Sodium carbonate	1. Potassium dichromate	1. Sodium oxalate	1. Sodium chloride
2. Oxalic acid	2. Borax	2. Potassium bromate	2. Potassium Ferro cyanide	2. Potassium chloride

Secondary Standard Solutions are those which are prepared by dissolving a little more than the gram equivalent weight of the substance per litre of the solution and then their exact standardization is done with primary standard solution. Some Secondary standards for acids are Sulphuric acid and Hydrochloric acid while for bases is Sodium hydroxide

Standard solutions of all the reagents required in a laboratory must be prepared and kept ready before taking up any analysis. However, their strength should be periodically checked or fresh reagents be prepared before analysis. In all titrations involving acidimetry and alkalimetry, standard solutions are required. These may be prepared either from standard substances by direct weighing or by standardizing a solution of approximate normality of materials by titrating against a prepared standard. The methods of preparation of standard solutions of some non-primary standard substances of common use are given below.

4.15.1 Standardization of Hydrochloric Acid (HCl):

The concentrated HCl is approximately 11N. Therefore, to prepare a standard solution, say deci-normal (0.1N) of the acid, it is diluted roughly one hundred times. Take 10 ml of acid and make approximately 1 litre by dilution with distilled water. Titrate this acid against 0.1N Na_2CO_3 (Primary standard) using methyl orange as indicator. Colour changes from pink to yellow when acid is neutralized. Suppose 10ml of acid and 12ml of Na_2CO_3 are consumed in the titration.

Acid Alkali

$$V_1 \times N_1 = V_2 \times N_2$$

$$10 \times N_1 = 12 \times 0.1$$

$$10 N_1 = 1.2$$

$$N_1 = 0.12$$

Normality of acid is 0.12. Similarly, normality of sulphuric acid can be worked out. H_2SO_4 needs to be diluted about 360 times to get approximately 0.1N because it has a normality of approximately 35. Then titrate against standard Na_2CO_3 to find out exact normality of H_2SO_4.

4.15.2 Standardization of Sodium Hydroxide (NaOH):

As per above method, the normality of HCl/H_2SO_4 has been fixed. Therefore, to find out the normality of sodium hydroxide, titration is carried out by using any one of these standard acids to determine the normality of the sodium hydroxide. For working out molarity, molar standard solutions are used. In case of the standardization of NaOH or any other alkali, potassium hydrogen phthalate can also be used as a primary standard instead of going through the titration with secondary standards. It can be decided depending upon the availability of chemicals in the laboratory.

Conversion Factors

In a laboratory exercise, various units of measurement, such as area, volume, mass, length, pressure, temperature etc. are required to be converted from SI to non-SI unit and vice-versa.

CHAPTER-5

Soil Sampling

Soil testing has become an important tool for assessing soil fertility and arriving at proper fertilizer recommendations. It's also a valuable management aid for studying soil changes resulting from cropping practices and for diagnosing specific cropping problems. Soil testing provides an index for the nutrient availability in soil and is a critical step in nutrient management planning. Soil sampling techniques, timing of sampling and type of analysis need to be considered for accurate results. The biggest problem in the effective use of soil testing is proper and representative sampling. Proper soil sampling will provide accurate soil test results and reliable nutrient recommendations. The following information is offered to answer special questions concerning soil sampling. Further information and guidance can be obtained by crop advisor.

5.1 When to Sample

Cultivated fields for spring seeding should be sampled after October. Fields for fall-seeded crops should be sampled one month before seeding. Forage fields for seed, pasture or hay may be sampled after September. Problem soil areas may be sampled any time. Frozen and water-logged soils should not be sampled because of the difficulty in obtaining a representative sample.

5.2 Types of Sampling

Fields can be broken into either zones or grids when developing a soil sampling plan. Within those zones or grids, soils can either be taken randomly or sampled at or near the intersections. Soil test values from random and grid sampling are often used to provide a single estimate for an entire field. This value may then be used to calculate fertilizer application rates.

5.2.1 Random Sampling

Uniform fields can be randomly sampled throughout the entire field. To see long-term trends in soil nutrient data, these points should be geo-referenced with a global positioning system (GPS) receiver and sampled in these same locations in subsequent years.

5.2.2 Grid Sampling

Grid sampling can be particularly useful where there is little prior knowledge of within-field variability. It also avoids sampling bias that could result from the collection of an unrepresentative composite sample due to a high portion of sub samples collected from the same region. Two common types of grid sampling include grid-cell and grid-point. Grid-cell soil sampling randomly collects either one or multiple sub samples throughout the cell for a composite sample. Grid-point soil sampling collects one or multiple sub samples around a geo-referenced point within a grid or at a grid intersection.

5.3 Types of Zone Sampling

Zone soil sampling technique that assumes each field contains different soils with unique soil properties and crop characteristics, and therefore should be separated into unique zones of management. For example, regions of fields that have had different crop history, yield or fertilizer treatments, and/or that vary substantially in slope, texture, depth and/or soil color should be separately sampled and therefore established as a zone. Unlike grid sampling, the number of zones and their shape and size will depend on the degree of field variability. In addition, zone sampling reduces the number of soil samples compared to grid or random sampling and allows for variable rate fertilizer applications ("prescription" rates). Variably applying fertilizer can improve yields, reduce fertilizer costs.

5.4 Soil Series

Soil series sampling identifies areas within and between fields that are unique from each other by using soil survey and topographic maps. Each soil series differs in its soil properties and will likely have different levels of available nutrients. Therefore, separate soil samples for each soil series in a field are collected. Soil test results may then be area-weighted based on the acreage of each soil series.

5.5 Topographic/Geographic Unit Sampling

Fields vary in natural features such as elevation, hilltops, slopes or depressions. Topographic/geographic unit sampling assumes these features differ in soil characteristics and therefore uses these features to establish unique zones. There are basically two different types of topographic/geographic unit sampling: area-based and point-based sampling. Area based soil sampling means that more than one soil sample is collected and composited from near the center of each topographic zone, whereas point-based soil sampling only collects one sample from the center of each topographic zone.

5.6 Remote Sensing Sampling

Remote sensing is the process of gathering data from a distance. It uses images collected by satellites or aircraft and combines those images with tabular information, digital maps and other digital data. That information is entered into a geographic information system (GIS), which is a computer database that retrieves, stores, analyzes and maps geographical information. The collected data or images, in the form of distinct wavelengths, are then formulated using common indices such as normalized difference vegetation index (NDVI), green normalized difference, vegetation index (GNDVI) or reflectance ratio vegetation index (RVI). The indices are mapped, indicating varying levels of a particular parameter such as plant nutrient content, water content, soil parameters (such as color) and yield. Because the relationship between indices and any of the above parameters are only estimates based on other research, calculated values should be ground-truthed and verified.

5.7 Yield Sampling

Crop growth and yields vary due to a number of soil parameters, such as texture, drainage, depth and management practices, including land shaping, spreader patterns and previous land use. Yield sampling zones use crop yield maps generated from combine yield monitor data, to determine where to soil sample. Yield data collected from yield monitors can be used in combination with GPS to map yields. Overall, yield maps are best used for zone delineation if the field is broken into arbitrary grids through a GIS program and the yields within each grid are averaged. Grids that have yields above the average are given a value of +1, yield grids below average are given a value of -1, and average yields for a grid are given a value of 0. If this procedure was repeated for each year's yield data, regardless of crop, a normalized yield frequency map would result when the multi-year normalized yield data were combined in a spreadsheet and then mapped. The resulting maps indicate zones that consistently yield high or low and those that do not. If a consistent factor controls yield variability in a field, then the distribution of this factor, and thus the distribution of crop yield, can assist in determining where to soil sample. For example, if low levels of a nutrient correspond to low yield areas, applying that nutrient should increase yield in those areas. However, if soil test results indicate adequate or high nutrient levels in low yielding areas, then the soil should be examined for compaction and other physical characteristics that could affect yield, particularly those that affect water storage or drainage. Fertilizer can then be reduced in these areas.

5.8 Management Zones

The management zone approach combines a number of zone sampling techniques to establish unique management zones. Combinations of prior experience, soil survey maps, yield maps, topography, electrical conductivity (EC; a measure of salinity) from sensors such as the Veris EC sensor or the EM-38 magnetic sensor, soil color, organic matter (O.M.), soil nutrients, moisture and remotely sensed vegetation indices are all useful in establishing multiple layers of information to develop unique zones. These layers of information may be used either by themselves (described above) or in other combinations to establish unique zones.

5.9 Recommendations Based on Research Results

Representative soil testing values of some soil nutrient have more spatial variability within a field than others. For example, phosphorus (P) levels have been observed to vary more than any other nutrient level within a field. The greatest variability is observed in areas with long cropping histories. For practical reasons, only one soil sampling strategy will generally be used for all tested nutrients; however, if one nutrient consistently limits yield, the method that is most accurate for that nutrient should be used. For example, area based topographic sampling is better than grid sampling at estimating nitrogen (N) concentrations (Franzen *et al.*, 1998). The grid approach is the best approach for measuring P in heavily fertilized fields, whereas both the grid and management zone approaches are good at measuring potassium (K) levels (Mallarino and Wittry, 2004). In addition, the grid-point method is better at measuring soil test P and K than the grid-cell method (Wollenhaupt *et al.* 1994). However, the management zone approach is the best approach for measuring O.M. and pH variability (Mallarino and Wittry, 2004). In areas with a history of lower soil P values or use of modest amounts of seed-placed starter fertilizer, a zone approach for all soil nutrients is valuable (Franzen, 2008). If a similar weight is given to all standard soil parameters, grid and management zone sampling should equally provide the greatest success at determining nutrient variability across all fields (Mallarino and Wittry, 2004). The management zone approach generally results in fewer soil samples than the grid approach, yet may take more planning time. The best strategy is to first determine the degree of variability within a field, and use grid sampling if variability is low (e.g. nutrient range is less than a factor of 2 to 3 across the field), and use zone sampling if variability is high.

The method and procedure for obtaining soil samples vary according to the purpose of sampling. Analysis of soil samples may be needed for engineering and agricultural purposes. In this publication, soil sampling for agricultural purpose is described which is done for soil fertility evaluation and fertilizer recommendations for crops. The results of even very carefully conducted soil analysis are as good as the soil sample itself. Thus, the efficiency of soil testing service depends upon the care and skill with which soil samples are collected. Non-representative samples constitute the largest single source of error in a soil fertility programme. It is to be noted that the most important phase of soil analysis is accomplished not in a laboratory but in the field where soils are sampled. Soils vary from place to place. In view of this, efforts should be made to take the samples in such a way that it is fully representative of the field. Only one to ten gram of soil is used for each chemical determination and represents as accurately as possible the entire surface 0-22 cm of soil, weighing about 2 million kg/ha.

5.10 Sampling Tools and Accessories

Depending upon the purpose and precision required, following tools may be needed for taking soil samples.

- Soil auger- it may be a tube auger, post hole or screw type auger or even a spade for taking samples.
- A clean bucket or a tray or a clean cloth for mixing the soil and sub sampling.
- Cloth bags of specific size.
- Copying pencil for markings and tags for tying cloth bags.
- Soil sample information sheet.

5.11 Selection of a Sampling Unit

A visual survey of the field should precede the actual sampling. Note the variation in slope, colour, texture, management and cropping pattern by traversing the field. Demarcate the field into uniform portions, each of which must be sampled separately. If all these conditions are similar, one field can be treated as a single sampling unit. Such unit should not exceed 1 to 2 ha, and it must be an area to which a farmer is willing to give separate attention. The unit of sampling is a compromise between the expenditure, labour and time on one hand and precision on the other. In view of limited soil testing facilities, it has been suggested to adopt an alternate approach where a sample may be collected from an area of 20-50 ha to be called as composite area soil sample and analyses the same for making a common recommendation for the whole area.

5.12 Sampling Procedure

Prepare a map of the area to be covered in a survey showing different sampling unit boundaries. A plan of the number of samples and manner of composite sampling is entered on the map, different fields being designated by letters A, B, C etc. Each area is traverse separately. A slice of the plough-layer is cut at intervals of 15 to 20 steps or according to the area to be covered. Generally 10 to 20 spots must be taken for one composite sample depending on the size of the field. Scrap away surface liter; obtain a uniformly thick slice of soil from the surface to the plough depth from each place. A V-shaped cut is made with a spade to remove 1 to 2 cm slice of soil. The sample may be collected on the blade of the spade and put in a clean bucket. In this way collect samples from all the spots marked for one sampling unit. In case of hard soil, samples are taken with the help of augur from the plough depth and collected in the bucket.

Pour the soil from the bucket on a piece of clean paper or cloth and mix thoroughly. Spread the soil evenly and divide it into 4 quarters. Reject two opposite quarters and mix the rest of the soil again. Repeat the process till left with about half kg of the soil, collect it and put in a clean cloth bag. Each bag should be properly marked to identify the sample. The bag used for sampling must always be clean and free from any contamination. If the same bag is to be used for second time, turn it inside out and remove the soil particles. Write the details of the sample in the information sheet. Put a copy of this information sheet in the bag. Tie the mouth of the bag carefully.

Precautions

- Do not sample unusual area like unevenly fertilized, marshy, old path, old channel, old bunds, area near the tree, site of previous compost piles and other unrepresentative sites.
- For a soft and moist soil, the tube auger or spade is considered satisfactory. For harder soil, a screw auger may be more convenient.
- Where crops have been planted in rows, collect samples from the middle of the rows so as to avoid the area where fertilizer has been band placed.

- Avoid any type of contamination at all stages. Soil samples should never be kept in the store along with fertilizer materials and detergents. Contamination is likely when the soil samples are spread out to dry in the vicinity of stored fertilizers or on floor where fertilizers were stored previously.
- Before putting soil samples in bags, they should be examined for cleanliness as well as for strength.
- Information sheet should be clearly written with copying pencil.

5.13 Dispatch of Soil Samples to the Laboratory

Before sending soil samples to the testing laboratory by a farmer, it should be ensured that proper identification marks are present on the sample bags as well as labels placed in the bags. It is essential that it should be written by copying pencil and not with ink because the ink will smudge and become illegible. The best way is to get the soil sampling bags from soil testing laboratory with most of the information printed or stenciled on them with indelible ink. Compare the number and details on the bag with the dispatch list. The serial numbers of different places should be distinguished by putting the identification mark specific for each centre. This may be in alphabets, say one for district and another for block and third for the village. Pack the samples properly. Wooden boxes are most suitable for long transport. Sample bags may be packed only in clean bags never used for fertilizer or detergent packing. Farmers may bring soil samples directly to the laboratory. Most of the samples are, however, sent to the laboratories through the field extension staff. An organized assembly-processing dispatch system is required to ensure prompt delivery of samples to the laboratory.

5.14 Preparation of Soil Samples for Analysis

5.14.1 Handling in the Laboratory

As soon as the samples are received at the soil testing laboratory, they should be checked with the accompanying information list. If the soil testing laboratory staffs have collected the samples themselves, then adequate field notes might have been kept. All unidentifiable samples should be discarded. Information regarding samples should be entered in a register and each sample be given a laboratory number, in addition to sample number, which helps to distinguish if more than one source of samples is involved.

5.14.2 Drying of Samples

Samples received in the laboratory may be moist. These should be dried in wooden or enameled trays. Care should be taken to maintain the identity of each sample at all stages of preparation. During drying, the trays can be numbered or a plastic tag could be attached. The soils are allowed to dry in the air. Alternatively, the trays may be placed in racks in a hot air cabinet whose temperature should not exceed 35^0 C and relative humidity should be between 30 and 60%. Oven drying a soil can cause profound change in the sample. This step is not recommended as a preparatory procedure in spite of its convenience. Drying has negligible effect on total N content but the nitrate content in the soil changes with time and temperature. Microbial population is affected due to drying at high temperature. With excessive drying, soil potassium may be released or fixed depending upon the original level of exchangeable potassium. Exchangeable potassium will be increased if its original level was less than 1 meq/100 g soil (1 cmol/kg) and *vice-versa*, but the effect depends upon the nature of clay minerals in the soil. In general, excessive drying, such as oven drying of the soil, affects the availability of most of the nutrients present in the sample and should be avoided. Only air drying is recommended. Nitrate, nitrite and ammonium determinations must be carried out on samples brought straight from the field. These samples should not be dried. However, the results are expressed on oven dry basis by separately estimating moisture content in the samples.

5.14.3 Post Drying Care

After drying, the samples are taken to the preparation room which is separate from the main laboratory. Air dried samples are ground with a wooden pestle and mortar so that the soil aggregate are crushed but the soil particles do not break down. Samples of heavy clay soils may have to be ground with an end runner grinding mill fitted with a pestle of hard wood and rubber lining to the mortar. Pebbles, concretions and stones should not be broken during grinding. After grinding, the soil is screened through a 2 mm sieve. The practice of passing only a portion of the ground sample through the sieve and discarding the remainder is erroneous. This introduces positive bias in the sample as the rejected part may include soil elements with differential fertility. The entire sample should, therefore, be passed through the sieve except for concretions and pebbles of more than 2 mm. The coarse portion on the sieve should be returned to the mortar for further grinding. Repeat sieving and grinding till all aggregate particles are fine enough to pass the sieve and only pebbles, organic residues and concretions remain out. If the soil is to be analyzed for trace elements, containers made of copper, zinc and brass must be avoided during grinding and handling. Sieves of different sizes can be obtained in stainless steel. Aluminum or plastic sieves are useful alternative for general purposes. After the sample is passed through the sieve, it must be again mixed thoroughly. The soil samples should be stored in cardboard boxes in wooden drawers. These boxes should be numbered and arranged in rows in the wooden drawers, which are in turn fitted in a cabinet in the soil sample room.

CHAPTER-6

Soil Physical Analysis

Soil physical measurements are numerous, depending on the objective of the study for agricultural purposes. These measurements generally include Soil water content, infiltration and hydraulic conductivity, evapotranspiration, heat, temperature reflectivity, porosity, particle size, bulk density, aggregate stability and particle size distribution. However, only a few physical measurements are normally conducted in soil-plant analysis laboratories.

6.1 Soil Water

As water is the most limiting factor in the agriculture, soil moisture determination is of major significance. Soil moisture influences crop growth not only by affecting plant physiological activities, but also acts as a solvent & nutrient carrier & maintains turgidity of the plants. In fact soil water is a regulator of physical, chemical & biological activities in the soil. The soil moisture is routinely measured in most field trials. While it can be assessed in the field by tensiometer, neutron moisture meter Gamma ray attenuation, time domain reflectometry techniques etc. however, the gravimetric approach is more flexible, as samples can be readily taken from any soil situation. All analyses in the laboratory are related to an air- or oven-dry basis, and therefore must consider the actual soil moisture content.

Apparatus

- Electric oven with thermostat.
- Desiccator.

Procedure

1. Weigh 10 g air-dry soil (< 2-mm) into a previously dried (105°C) and weighed metal Can with lid.
2. Dry in an oven, with the lid unfitted, at 105°C overnight.

3. Next day, remove from oven; fit the lid, cool in a desiccator for at least 30 minutes and re-weigh.

Calculations

$$\% \, Moisture \, in \, soil \, (Q) = \frac{Wet \, soil(g) - Dry \, soil \, (g)}{Dry \, soil \, (g)} \times 100$$

6.2 Particle Size Distribution

Individual soil particles vary widely in any soil type. Similarly, as these particles are cemented together, a variety of aggregate shapes and sizes occur. For standard particle size measurement, the soil fraction that passes a 2-mm sieve is considered. Laboratory procedures normally estimate percentage of sand (0.05 - 2.0 mm), Silt (0.002 - 0.05 mm), and Clay (<0.002 mm) fractions in soils. Particle size distribution is an important parameter in soil classification and has implications on soil water, aeration, and nutrient availability to plants. As primary soil particles are usually cemented together by organic matter, this has to be removed by H_2O_2 treatment. However, if substantial amounts of $CaCO_3$ are present, actual percentages of sand, silt or clay can only be determined by prior dissolution of the $CaCO_3$. The two common procedures used for particle size analysis or mechanical analysis are the bouyoucos hydrometer method or the pipette gravimetric method. The hydrometer method of silt and clay measurement relies on the effect of particle size on the differential settling velocities within a water column. Theoretically, the particles are assumed to be spherical having a specific gravity of 2.65 g/cm^3. If all other factors are constant, then the settling velocity is proportional to the square of the radius of the particle (Stoke's Law). The settling velocity is also a function of liquid temperature, viscosity and specific gravity of the falling particle. In practice, therefore, we must know and make corrections for the temperature of the liquid. Greater temperatures result in reduced viscosity, due to liquid expansion and a more rapid descent of falling particles.

Apparatus

Soil dispersing stirrer: A high-speed electric stirrer with a cup receptacle. Hydrometer with Bouyoucos scale in g/ltr.

Reagents

a. Dispersing agent

Dissolve 40 g sodium hexametaphosphate [$(NaPO_3)_{13}$], and 10 g sodium carbonate (Na_2CO_3) in distilled water, and bring to 1-ltr. volume with distilled water. This solution deteriorates with time and should not be kept for more than 1 to 2 weeks.

b. Amyl alcohol

Procedure

1. Weigh 40 g air-dry soil (2-mm) into a 600-mL beaker.
2. Add 60-ml dispersing solution.
3. Cover the beaker with a watch-glass, and leave overnight.
4. Quantitatively transfer contents of the beaker to a soil-stirring cup, and fill the cup to about three-quarters with water.
5. Stir suspension at high speed for 3 minutes using the special stirrer. Shake the suspension overnight if no stirrer is available.
6. Rinse stirring paddle into a cup, and allow standing for 1 minute.
7. Transfer suspension quantitatively into a 1-litre calibrated cylinder (hydrometer jar), and bring to volume with water.

a. Determination of Blank

Dilute 60 ml dispersing solution to 1-litre hydrometer jar with water.

- Mix well, and insert hydrometer, and take hydrometer reading, (R_b)
- The blank reading must be re-determined for temperature changes of more than 2°C from 20°C.

b. Determination of Silt plus Clay

- Mix suspension in the hydrometer jar, using a special paddle carefully, withdraw the paddle, and immediately insert the hydrometer.
- Disperse any froth, if needed, with one drop of amyl alcohol and take hydrometer reading 40 seconds after withdrawing the paddle. This gives reading, (R_{SC}) and percentage [silt + clay] in soil can be calculated as:

$$\% \, [Silt + Clay] \left(\frac{w}{w}\right) = (R_{sc} - R_b) x \frac{100}{Oven\ dry\ soil\ (g)}$$

c. Determination of clay

- Mix suspension in the hydrometer jar with paddle, withdraw the paddle, with great care, leaving the suspension undisturbed.
- After 4 hours, insert the hydrometer, and take hydrometer reading, (R_c) and percentage [clay] in soil can be calculated as:

$$\% \, Clay \left(\frac{w}{w}\right) = R_c - R_b \, x \frac{100}{Oven\ dry\ soil\ (g)}$$

Thus, percentage silt in soil can be calculated as:

$$\% \, Silt \left(\frac{w}{w}\right) \left[\% \, Silt + Clay \left(\frac{w}{w}\right)\right] - \left[\% \, Clay \left(\frac{w}{w}\right)\right]$$

d. Determination of Sand

- After taking readings required for clay and silt, pour suspension quantitatively through a 50-µm sieve.
- Wash sieve until water passing the sieve is clear.
- Transfer the sand quantitatively from sieve to a 50 ml beaker of known weight.
- Allow the sand in the beaker to settle, and decant excess water.
- Dry beaker with sand overnight at 105°C.
- Cool in a desiccator, and re-weigh beaker with sand.

Then, Percentage Sand in soil can be calculated as:

$$\% \, Sand \left(\frac{w}{w}\right) = Sand \; weight \; x \frac{100}{Oven \; dry \; soil \; (g)}$$

Where weight of sand follows from

$$Sand \; weight \; (g) = [Beaker + Sand \; (g)] - [Beaker \; (g)]$$

Note

1. If possible, all hydrometer jars should be placed in a water bath at constant temperature (20°C); in that case, temperature corrections are not needed.
2. For temperature correction, use a value of 0.4 for each degree temperature difference from 20°C. Add or subtract this factor if the temperature is more or less than 20°C, respectively.
3. All results of mechanical analysis should be expressed on the basis of oven dry soil (24 hours drying at 105°C).
4. In the above procedure, carbonates and organic matter are not removed from the soil.
5. The Hydrometer method, as described in this section, cannot be applied to soils that contain free gypsum (gypsiferous soils). For gypsiferous soils, see Hesse (1971).
6. Sum of % silt and clay + % sand should be 100 %. The magnitude of deviation from 100 is an indication for the degree in accuracy.

6.2.1 Soil Texture

Once the percentage of sand, silt, and clay is measured, the soil may be assigned a textural class using the USDA textural triangle Fig. 6.1 Within the textural triangle are various soil textures which depend on the relative proportions of the soil fractions.

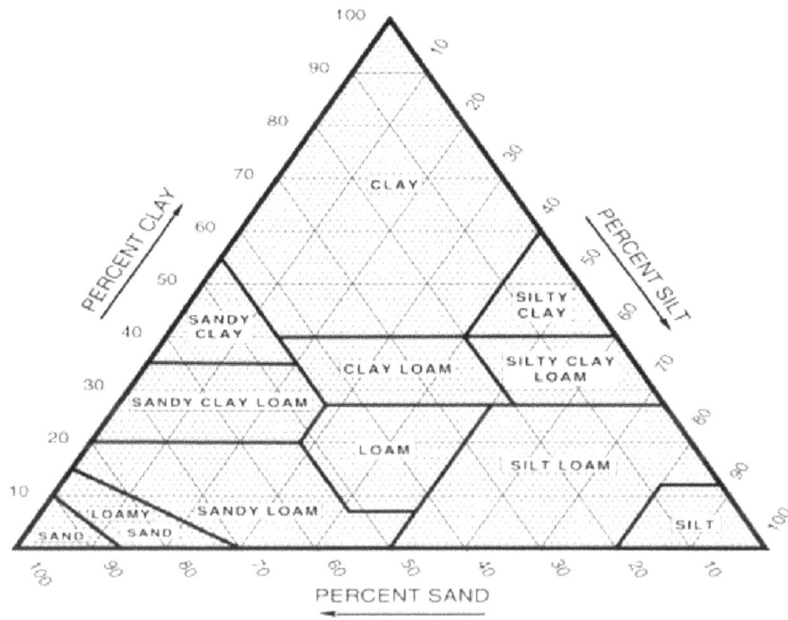

Figure 6.1 : USDA Textural triangle for determination of texture of soils

6.3 Soil Structure

Soil structure is defined as the arrangement of the soil particles. With regard to structure, soil particles refer not only to sand, silt and clay but also to the aggregate or structural elements, which have been formed by the aggregation of smaller mechanical fractions. The word 'particle' therefore, refers to any unit that is part of the make-up of the soil, whether primary unit (sand, silt or clay fraction) or a secondary (aggregate) particle. The size, shape and character of the soil structure varies, which could be cube like, prism like or platter like. On the basis of size, the soil structure is classified as follows:

- Very coarse: >10 mm
- Coarse: 5-10 mm
- Medium: 2-5 mm
- Fine: 1-2 mm
- Very fine: <1 mm

Depending upon the stability of the aggregate and the ease of separation, the structure is characterized as follows:

- Poorly developed
- Weekly developed

- Moderately developed
- Well developed
- Highly developed

The soil structure or aggregate consists of an intermediate grouping of number of primary particles into a secondary unit. The important factors which facilitate the aggregation of soil particles are:

- Clay particles and types of clay minerals
- Cations such as calcium
- Organic matter
- Colloidal matter such as oxides of iron and aluminium
- Plant roots
- Soil microbes and their types where fungi being most effective

Soil structure influences the extent of pore space in the soil, water holding capacity, aeration, root movement and the nutrient availability. The better and more stable soil aggregates are considered as a desirable soil property with regard to plant growth. Determination of soil structure is, therefore, an important exercise in soil fertility evaluation programme. An aggregate analysis aims to measure the percentage of water stable secondary particles in the soil and the extent to which the finer mechanical separates are aggregated into coarser fractions. The determination of aggregate or clod size distribution involves procedures that depend on the disintegration of soil into clods and aggregates. The resulting aggregate size distribution depends on the manner and condition in which the disintegration is brought about. For the measurements to have practical significance, the disruptive forces causing disintegration should closely compare with the forces expected in the field. The field condition particularly with respect to soil moisture should be compared with the moisture condition adopted for soil disintegration in the laboratory. The sampling of soil and subsequent disintegration of clods in relevance to seed bed preparation for upland crops should be carried out under air dry conditions for dry sieve analysis. A rotary sieve shaker would be ideal for dry sieving. Similarly the processes of wetting, disruption of dry aggregates and screening of aggregates should be compared to the disruptive actions of water and mechanical forces of tillage under wetland conditions. Although vacuum wetting of dry soil largely simulates the process of wetting in-situ, particularly in the subsurface layers, the surface soil clods, however, experience large scale disruption when they are immersed in water at atmospheric pressure. The re-productivity of the size distribution of clods should naturally be the criterion for deciding the method of wetting by either vacuum wetting or immersion in water. The immersion wetting is more closer to wetting of surface soil by irrigation. After wetting, aggregates of different sizes can be obtained through several methods like sedimentation, elutriation and sieving. However, sieving under water compares more closely with the disruptive actions of water and other mechanical forces as experienced during wetland rice field preparation.

6.3.1 Dry Aggregates Analysis (Gupta and Ghil Dyal, 1998)

The size distribution of dry clods is measured by Dry Sieving Analysis performed on air dry bulk soil sample either manually or with the help of a rotary sieve shaker.

Apparatus

- Nest of sieves, 20 cm in diameter and 5 cm in height, with screens having 25.0, 10.0, 5.0, 2.0, 1.0, 0.5 and 0.25 mm size round openings with a pan and a lid
- Rotary sieve shaker
- Aluminum cans
- Balance
- Spade
- Brush
- Polyethylene bags
- Labels

Procedure

Bulk soil sample is collected from the tilled field with the help of a 20 cm diameter and 10 cm height ring. The ring is placed on the tilled soil and pressed until in level with the surface. The loose soil within the ring is removed and collected in a polyethylene bag. One label indicating the depth and soil sample is put inside the bag and the other label is tied with the bag. The soil samples are then brought to the laboratory and air dried. Spread the soil on a sheet of paper and prepare the sub samples by 'quartering'. The mixed soil material is coned in the center of the mixing sheet with care to make it symmetrical with respect to fine and coarse soil material. The cone is flattened and divided through the center with a flat metal spatula or metal sheet, one-half being moved to the side quantitatively. Each one-half is further divided into halves; the four quarters being separated into separate piles or 'quarters'. The sub-samples from two of these 'quarters' are weighed and used for clod size and aggregate distribution analysis as duplicates. The weighed soil sample is transferred to the top sieve of the nest of sieves having 5.0, 2.0, 1.0, 0.5 and 0.25 mm diameter round openings and a pan at the bottom. Cover the top sieve with the lid and place the nest of sieves on a rotary shaker. Switch on the shaker for 10 minutes, and then remove the sieves, and collect the soil retained on each screen in the pre-weighed aluminum cans, with the help of a small brush, and weigh the cans with the soil. If the percentage of dry aggregates on 5 mm sieve exceeds 25%, transfer these aggregates to a nest of sieves with 25.0, 10.0 and 5.0 mm sieves along with a pan. Cover the top sieve containing the aggregates with a lid and place the nest of sieves on the rotary sieve shaker. Switch on the motor for 10 minutes and proceed as above for the estimation of aggregate size distribution. Analyse the duplicate sample following the same procedure and calculate the percent distribution of dry aggregates retained on each sieve. Duplicate 100 g sample is dried in an oven for 24 hours at 105^0 C to calculate the oven dry weight of the soil sample.

Calculation

i. Weight of Aggregates in each sieve group = (Wt. of Aggregates + Can) – Wt.of Can

ii. Percent distribution of Aggregates in each size group =

$$\frac{\text{Weight of Aggregate in each size group } \times 100}{\text{Total weight of soil}}$$

$$\text{Oven-dry weight of Aggregate (\%)} = \frac{\text{Air-dry weight (\%) } \times 100}{100 + \text{Moisture\%}}$$

6.3.2 Wet Aggregate Analysis (Gupta and Ghil Dyal, 1998)

Apparatus

- A mechanical oscillator powered by a gear reduction motor having amplitude of oscillation 3.8 cm and frequency of oscillator 30-35 cycles per minute
- Two sets of sieves, each having 20 cm diameter and 5 cm height with screen openings of 5.0, 2.0, 1.0, 0.5, 0.25 and 0.1 mm diameter
- Two Buckner funnels 15 cm in diameter with rubber stoppers
- Two vacuum flasks of one litre capacity
- Suction pump or aspirator
- Rubber policeman
- Twelve aluminium cans
- Perforated cans
- Sand bath
- Filter papers

Reagents

- 5% Sodium hexametaphosphate
- 4% Sodium hydroxide

Procedure

Among the different procedures adopted, wetting the samples under vaccum is suggested because the rate of wetting influences slaking of crumbs. The time of sieving ranges from 10 to 30 minutes depending upon the type of wetting. Ten minutes pre-shaking of the soil sample in a reciprocating shaker or end-to-end shaker has been suggested by Baver and Rhodes (1932) for fine textured soil .The technique used by Yoder (1936) and subsequently improved by the Soil Science Society of

America's Committee on Physical Analysis is generally used for determining the size distribution of water stable aggregates.

1. The soil sample is taken when it is moist and friable. It is broken by applying mild stress into smaller aggregates which can pass through 8 mm screen.
2. The sieved soil sample is taken on a watch glass for wetting by either vacuum soaking or immersion method.
3. For vacuum wetting, the sample is placed in a vacuum desiccator containing de-aerated water at the bottom. The desiccator is evacuated until the pressure inside drops to about 3 mm and water starts boiling.
4. Water is then allowed to enter through the top of the desiccator and to flow into the watch-glass holding the sample. Enough water is added to cover the soil sample. Soil sample is then taken out of desiccator.
5. Prepare four soil samples of 25 g each, Place a set of duplicate sample in an oven for the determination of moisture content. Another set of saturated duplicate soil sample is then transferred to the top sieve of the nest of sieves (5.0, 2.0, 1.0, 0.5, 0.25 and 0.1 mm) and spread with the help of a glass rod and a slow jet of water.
6. The bottom pan is then removed, and the nest of sieves is attached to the Yoder type wet sieve shaker. Fill the drum (which holds the set of sieves) with salt-free water at 20-25^0 C to a level somewhat below that of the screen in the top sieve of the nest of sieves, when the sieves are in the highest position.
7. Then lower the nest of sieves to wet the soil for 10 minutes. Bring the nest of sieves to the initial position and adjust the level of water so that the screen in the top sieve is covered with water in its highest position.
8. Now switch on the mechanical oscillator to move the nest of sieves up and down with a frequency of 30-35 cycles per minute and a stroke of 3.8 cm. Sieving is done for 10 minutes. Remove the nest of screens from the water and allow it to drain for some time.
9. Transfer the soil resting on each screen with a stream of distilled water and brush into a Buckner funnel having a pre-weighed filter paper and connected to a suction pump. Transfer the soil along with the filter paper into an aluminium can and dry at 105^0 C for 24 hours.
10. Weigh the soil nearest to 0.01 g. Transfer the oven dry soil aggregates from all the cans of a set into the dispersion cup. Add dispersing agent (10 ml of 5% solution of sodium hexametaphosphate for normal and Ca-saturated soils, or 10 ml of 4% solution of sodium hydroxide for acid soil) and enough distilled water to fill the cup within 4 cm of the rim, and then stir the suspension for 10 minutes.
11. Wash the suspension on an identical set of sieves as used previously by means of a stream of tap water and a brush and transfer it to aluminium cans. The sand in each can is oven dried and weighed in the same manner as above.

12. Calculate the percent distribution of soil particles (aggregates and the sand) and the sand particles retained on each sieve.

Calculation

Size distribution of soil particles (aggregate + sand):

$$\text{Soil particles in each size group (\%)} = \frac{W_{\text{oven dry}}(ag+s)i \times 100}{W_{oven\,dry}}$$

$$\text{Sand particle in each size group (\%)} = \frac{W(s) \times 100}{W_{oven\,dry}}$$

Where,

$W_{\text{oven dry}}$ is the oven dried weight of the aggregates (ag) and Sand (s), and "i" is the size group.

6.4 Soil Moisture

Soils are equilibrated with water at various tensions and moisture content is determined. The ability of soil to retain water depends on several factors, e.g., texture or particle-size distribution, organic matter content (due to its hydrophilic nature),nature of mineral colloids, and soil structure or arrangement of particles.

6.4.1 Low Range at 0-100 Kpa (0-10 bar) Pressure

Apparatus

- One-bar pressure plate extractor
- One-bar ceramic plates.
- Rubber rings (5-cm diameter, 1-cm height).
- Compressed air source with a manifold, regulator, and gauge.
- Balance.
- Drying oven.
- Disposable aluminum dishes or soil-moisture cans.
- Desiccator.

Procedure

1. Submerge the ceramic plates in water for 24 hours to saturate.
2. Place plates on a workbench.
3. Place labeled rubber rings in order on the plate (each plate accommodates 12 samples).
4. Fill ring with 2-mm air-dry soil using a spatula (about 20 g sample). In order to avoid particle-size segregation, place entire soil sample into the ring.

5. Level, but don't pack, the sample in the ring.
6. Cover plate with water to wet sample from below. Add water between the rings until there is an excess of water (at least 3-mm deep) on the plate.
7. Cover samples with wax paper or a plastic sheet.
8. Allow samples to stand overnight.
9. The next morning, remove excess water from the plate with a syringe, disposable pipette, or siphon.
10. Place the triangular support in the extractor vessel on the bottom.
11. Install plate with samples in the lower-most position in the extractor. Then install the middle and top plates (the plastic spaces should be placed between plates).
12. Connect outflow tubes.
13. Close extractor and tighten, ensuring that the "O" ring is in place and all nuts are uniformly tightened. Apply desired pressure in the 0 - 100 kpa (0 - 1 bar) range. Build up the pressure in the vessel gradually.
14. Place a beaker to collect water from the outflow tubes.
15. Maintain pressure until no more water is being released (generally 18 – 20 hours, but for some soils 48 hours or even longer).
16. Release pressure from extractor (remove outflow tubes from water before turning instrument off).
17. Open extractor.
18. Without undue delay, transfer moist soil sample from ring with a wide-bladed spatula to a tarred dish. (It is not necessary to make a quantitative transfer of the entire soil.)
19. Immediately weigh wet sample (accuracy 0.01 g) and place in drying oven at105 °C for 24 hours.
20. Place sample in desiccator, cool, and weigh.

Calculation

$$\%\,Moisture\,(\theta) = \frac{Wet\ soil\,(g) - Dry\ Soil\,(g)}{Dry\ soil\,(g)} x\ 100$$

6.4.2 High Range Moisture at 100-1500 k pa (0-15 bar) Pressure

Apparatus

Fifteen-bar ceramic plate extractor.

Fifteen-bar ceramic plates.

Rubber rings.

Balance.

Drying oven.

Weighing dishes (disposable aluminum dishes or tarred soil-moisture cans)

Burette.

Desiccator.

Procedure

1. Use 15-bar ceramic plates and follow Steps 1 - 12 of the previous method(6.4.1). Applying 1 - 15 bar pressure (100 - 1500 kpa).
2. Place beaker to collect water from outflow tubes.
3. Leave overnight.
4. Connect outflow tube to burette partially filled with tap water.
5. Samples should stay in extractor until flow has ceased from all samples on plate and the soils have reached equilibrium (24 - 48 hours for most soil; however some fine textured and organic soils may needs up to 120 hours).

   No change in reading on burette would indicate that flow has stopped from all samples and equilibrium has been attained.
6. Disconnect burette to prevent back flow of tap water.
7. Release pressure from extractor.
8. Follow Steps 17 - 20 of the previous method(6.4.1).

Calculation

$$\% \, Moisture \, (\theta) = \frac{Wet \, soil \, (g) - Dry \, Soil \, (g)}{Dry \, soil \, (g)} x \, 100$$

1. If the outlets of the plates continue to bubble after a few hours of applied pressure, the plates are probably defective and should be replaced.
2. Pressure should not be allowed to fluctuate during a run. It should be checked after every 2 - 3 hours (and adjusted if necessary). If the pressure fluctuations are within the specified tolerance of the regulator, then no adjusting is needed.
3. Never remove extractor lid with pressure in the container.
4. The height of the sample in the ring should be as small as possible to reduce the time required to reach equilibrium, which is proportional to the square of the height of the sample in the ring.

6.4.3 Available Water Capacity Estimation

Available water capacity is the amount of water retained in the soil reservoir that can be removed by plants. It is estimated by the difference in the soil water content between **Field Capacity** and **Permanent Wilting Capacity**

$$\% \, AWC = FC \, (\%) - PWD \, (\%)$$

6.4.4 Field Capacity Estimation

Field capacity is commonly estimated by measuring the moisture retained at the following pressure:

Coarse-textured Soils 10 Kpa(1/10 bar)

Medium-textured Soils 33 Kpa (1/3 bar)

Fine-textured Soils 50 Kpa (1/2 bar)

6.4.5 Permanent Wilting Point Estimation

Wilting point is commonly estimated by measuring the 1500-kpa (15-bar) percentage. It varies according to plant species and stage of plant growth, ranging 10 to 25 bars for mesophytic plants.

6.4.6 Water Holding Capacity

Field capacity or water holding capacity as the amount of water held in the soil after the excess gravitational water has drained away and after the rate of downward movement of water has materially ceased. Stage of field capacity is attained in the field after 48 to 72 hours of saturation. It is the upper limit of plant available soil moisture.

Apparatus

- Polythene sheets
- Spade
- Soil auger
- Moisture boxes/cans
- Balance
- Oven
- Oven - dry weight of soil

Procedure

1. Select a uniform plot measuring 5 m x 5 m.
2. Remove weeds, pebbles etc. and make bunds around the plot.
3. Fill sufficient water in the plot to completely saturate the soil.
4. Cover the plot area with a polythene sheet to check evaporation.
5. Take soil sample from the centre of the plot from the desired layer, starting after 24 hours of saturation and determine moisture content daily till the values of successive days are nearly equal.
6. Record the weight as below:

Weight of empty moisture box = X

Weight of moisture box + moist soil = Y

Weight of moisture box + oven dry soil = Z

Repeat above on next day and so on till a constant Z value is reached.

Calculation

Moisture content in soil = Y – Z

Weight of oven dry soil = Z – X

$$\text{Percentage of moisture in soil (1st day)} = \left(\frac{Y-Z}{Z-X}\right) \times 1000 = a$$

Percentage of moisture on succeeding days = a1, a2, etc.

Plot the daily readings on a graph paper. The lowest reading is taken as a value of field capacity of the soil.

CHAPTER-7

Soil Chemical Properties

7.1 Cation Exchange Capacity (CEC)

The total number of exchangeable cations a soil can hold is called cation exchange capacity (CEC). The higher the CEC, the more cations it can retain. It can be expressed in terms of milli-equivalents/100 g of soil (meq/100 g) or centimoles of positive charge per kg of soil cmol(p^+)kg^{-1}, which is numerically equal to meq/100 g. The CEC of the soil depends on the kind of clay and organic matter present.

Apparatus

- Centrifuge
- 50 ml round bottom centrifuge tubes
- Mechanical shaker
- Flame Photometer and accessories which include Propane, Lithium and Sodium standards

Reagents

- Sodium acetate (NaOAc) 1.0M: Dissolve 136.08 g of sodium acetate trihydrate in distilled water and bring the volume to 1 litre. Adjust the pH to about 8.2.
- Ethanol 95%.
- Ammonium acetate (NH_4OAc) 1.0M: Dissolve 77.09 g of ammonium acetate in distilled water and dilute to approx. 900 ml. Adjust pH to 7.0 with dilute ammonium hydroxide or acetic acid as required and make up the volume to 1litre.
- Standard solution of NaCl: Dissolve 5.845 g of AR grade NaCl in 1.0M ammonium acetate and make the volume to 1 litre. It will give 100 meq/ litre of sodium in stock solution. From this solution take 0, 1, 2, 5, 7.5 and 10

ml and make up the volume to 100 ml each with the ammonium acetate. It will give 0, 1, 2, 5, 7.5 and 10 meq/litre of sodium.

Procedure

1. Weigh accurately 5 g soil and transfer the sample to a 50 ml centrifuge tube.
2. Add 25 ml of 1.0M sodium acetate solution to the tube stopper and shake in a mechanical shaker for 5 minutes.
3. Centrifuge at 2000 rpm for 5 minutes or until the supernatant liquid is clear.
4. Decant the liquid completely and repeat the extraction three more times. Discard the decants.
5. Repeat steps 2 – 4 with ethanol or isopropyl alcohol until the EC of the decant reads less than 40 mS/cm (usually it takes 4 to 5 washings).
6. To displace the adsorbed Na, repeat steps 2 – 4 using the ammonium acetate solution. Collect the decant in 100 ml volumetric flask fitted with a funnel and filter paper. Make up to volume with ammonium acetate solution.
7. To determine sodium concentration by flame photometry, prepare a series of Na standard solutions in the range of 0 – 10 meq/litre of Na. Prepare a standard curve by plotting sodium concentration on x-axis and flame photometric readings on y-axis. Unknown sample extract is fed on the flamephotometer and the reading is taken, corresponding to which the concentration of sodium is read from the standard curve. For better results, add LiCl in each standard to yield a final concentration of about 5 meq/litre of LiCl.

Calculation

Ammonium acetate extractable Na which is exchangeable Na in meq/100 g soil =

$$\frac{\text{Na. Conc. of extract in meq/litre(Y)}}{\text{Wt. of soil in g(5)}} \times \frac{100 \times \text{Vol. of extract in ml (100)}}{1000} = \frac{Y \times 10}{5} = 2Y$$

This displaced Na is actually a measure of the Cation Exchange Capacity (CEC) of the soil. So, meq Na/100 g soil is actually meq exchangeable cations (Ca, Mg, Na, and K)/100 g soil.

7.2 Soil Reaction [pH]

The soil pH is the negative logarithm of the active hydrogen ion (H^+) conc. in the soil solution. It is the measure of soil sodicity, acidity or neutrality. It is a simple but very important estimation for soils, since soil pH influences to a great extent the availability of nutrients to crops. It also affects microbial population in soils. Most nutrient elements are available in the pH range of 5.5 to 6.5. In various chemical estimations, pH regulation is critical. Specific colours as observed in the presence of various pH indicators and the colour changes due to pH change. The procedure for measurement of soil pH is given below.

Apparatus

- pH meter with a range of 0-14 pH
- Pipette/dispenser
- Beaker
- Glass rod

Reagent

- Buffer solutions of pH 4, 7 and 9: Dissolve seperately one tablet of each buffer in water and make the volume to 100ml.
- Calcium chloride solution (0.01M): Dissolve 1.47 g $CaCl_2.2H_2O$ in 1 litre of water to obtain 0.01M solution.

Procedure

1. Calibrate the pH meter, using 2 buffer solutions, one should be the buffer with neutral pH (7.0) and the other should be chosen based on the range of pH in the soil. Take the buffer solution in the beaker. Insert the electrode alternately in the beakers containing 2 buffer solutions and adjust the pH. The instrument indicating pH as per the buffers is ready to test the samples
2. Weigh 10 g of soil sample into 50 or 100 ml beaker, add 20ml of $CaCl_2$ solution (use water instead of $CaCl_2$ solution throughout the procedure if water is used as a suspension medium).
3. Allow the soil to absorb $CaCl_2$ solution without stirring, and then thoroughly stir for 10 second using a glass rod.
4. Stir the suspension for 30 minutes and record the pH on the calibrated pH meter.

Based on soil pH values, following types of soil reactions are distinguished:

Soil Reaction Rating

<4.6	-	Extremely acid
4.6-5.5	-	Strongly acid
5.6-6.5	-	Moderately acid
6.6-6.9	-	Slightly acid
7.0	-	Neutral
7.1-8.5	-	Moderately alkaline
>8.5	-	strongly alkaline

The acidic soils need to be limed before they can be put to normal agricultural production. The alkali soils need to be treated with gypsum to remove the excessive content of sodium.

7.2.1 Lime Requirement

Crop yields are normally high in soils with pH values between 6.0 and 7.5. Lime is added to raise the pH of acid soils, and the amount of lime required to raise the pH to an optimum level is called as Lime Requirement. A number of methods are available for the determination of lime requirement. The Woodruff and the Shoemaker *et al* methods are discussed here which are based on the use of a buffer solution, whose pH undergoes change when treated with acid soils. The pH of buffer solution will gradually decrease when H^+ ion concentration increases. When H^+ increases by 1 meq in 100 ml buffer solution, pH value will decrease by 0.1 unit. Buffer solutions needs to be prepared a fresh. A 0.05M solution of AR grade potassium hydrogen pthalate (molecular weight 204.22) gives a pH of 4.0 at 25 ^{0}Cand it can be used as a buffer.

7.2.1.1 Woodruff's Method (Woodruff, 1948)

Apparatus

- pH meter
- Automatic pipettes

Reagent

Woodruff's buffer solution: Dissolve 10 g calcium acetate [$Ca(CH_3COOH)_2$], 12g para-nitrophenol, 10 g salicylic acid and 1.2 g sodium hydroxide in distilled water. Adjust pH to 7.0 with acetic acid or sodium hydroxide, transfer to 1 litre volumetric flask and make the volume to the mark with distilled water.

Procedure

1. Take 10 g soil sample in a clean 50 ml beaker.
2. Add 10 ml of distilled water, stir and wait for 30 minutes.
3. Determine pH value in soil suspension.
4. If the pH value is less than 5.0 (average of 4.5 and 5.5 to have one value), add 10ml Woodruff's buffer solution, stir and wait for 30 minutes before determining new pH value.

The soils between pH 6.6 and 7.5 are practically considered as nearly neutral. Such soils do not need to be treated with lime or gypsum. Even in case of soils, which are acidic and alkaline beyond these limits, growing of acid loving and salt tolerant crops may be considered. Only highly acidic soils and the soils with high alkalinity need to be treated with chemical amendments since this operation is quite expensive.

7.2.1.2 Shoemaker Method (Shoemaker *et al.*, 1961)

Apparatus

- pH meter
- Automatic pipettes - 10 and 20 ml

Reagent

- Extractant Buffer: Dissolve 1.8 g of nitrophenol, 2.5 ml triethanolamine, 3.0 g potassium chromate (K_2CrO_4), 2.0 g calcium acetate and 53.1 g calcium chloride in a litre of water. Adjust pH to 7.5 with NaOH.

Procedure

1. Take 5.0 g soil sample in a 50 ml beaker.
2. Add 5 ml distilled water and 10 ml extractant buffer.
3. Shake continuously for 10 minutes or intermittently for 20 minutes and read the pH of the soil buffer suspension with glass electrode. The pH of the buffer solution is reduced, depending upon the extent of soil acidity. For various levels of measured pH of soil buffer suspension, the amount of lime required to raise the soil pH to 6.0, 6.4 and 6.8 is given in Table 7.1 in terms of $CaCO_3$.The decrease of buffer pH by 0.1 unit is equivalent to 1 meq of H^+ in 100 ml buffer solution. The lime requirement varies with the type of soils and their cation exchange capacity.

Table 7.1 Lime requirement for different pH targets

Measured pH of	Lime Requirement in tonnes/ha as $CaCO_3$ for bringing soil pH to different levels of soil buffer suspension		
ph	6.0	6.4	6.8
6.7	2.43	2.92	3.40
6.6	3.40	4.13	4.62
6.5	4.37	5.35	6.07
6.4	5.59	6.56	7.53
6.3	6.65	7.78	8.99
6.2	7.52	8.93	10.20
6.1	8.5	10.21	11.66
6.0	9.48	11.42	13.12

Practically, pH of acid soils may not be raised beyond 6.4/6.5.

7.3 Electrical Conductivity (EC)

The electrical conductivity (EC) is a measure of the ionic transport in a solution between the anode and cathode. This means, the EC is normally considered to be a measurement of the dissolved salts in a solution. Like a metallic conductor, they obey Ohm's law. Since the EC depends on the number of ions in the solution, it is important to know the soil/water ratio used. The EC of a soil is conventionally based on the measurement of the EC in the soil solution extract from a saturated soil paste, as it has been found that the ratio of the soil solution in saturated soil paste is approximately two-three times higher than that at field capacity. As the determination of EC of soil solution from a saturated soil paste is cumbersome and demands 400-500 g soil sample for the determination, a less complex method is normally used. Generally a 1:2 soil/water suspension is used.

- EC meter
- Beakers (25 ml), erlenmeyer flasks (250 ml) and pipettes
- Filter paper

Reagent

- 0.01M Potassium chloride solution: Dry a small quantity of AR grade potassium chloride at 60 ^{0}C for two hours. Weight 0.7456 g of it and dissolve in freshly prepared distilled water and make the volume to one litre. This solution gives an electrical conductivity of 1411.8x10^{-3} i.e. 1.412 mS/cm at 25^{0}C. For best result, select a conductivity standard (KCl solution) close to the sample value.

Procedure

1. Take 40 g soil into 250 ml Erlenmeyer flask, add 80 ml of distilled water, stopper the flask and shake on reciprocating shaker for one hour. Filter through Whatman No.1 filter paper. The filtrate is ready for measurement of conductivity.
2. Wash the conductivity electrode with distilled water and rinse with standard KCl solution.
3. Pour some KCl solution into a 25 ml beaker and dip the electrode in the solution. Adjust the conductivity meter to read 1.412 mS/cm, corrected to 25^0 C.
4. Wash the electrode and dip it in the soil extract.
5. Record the digital display corrected to 25^{0}C. The reading in mS/cm of electrical conductivity is a measure of the soluble salt content in the extract, and an indication of salinity status of this soil. The conductivity can also be expressed as mmhos/cm.

7.4 Gypsum Requirement (Schoonover, 1952)

In the estimation of gypsum requirement of saline-sodic/sodic soils, the attempt is to measure the quantity of gypsum (Calcium sulphate) required to replace the sodium from the exchange complex. The sodium so replaced with calcium of gypsum is removed through leaching of the soil. The soils treated with gypsum become dominated with calcium in the exchange complex. When Calcium of the gypsum is exchanged with sodium, there is reduction in the calcium concentration in the solution. The quantity of calcium reduced is equivalent to the calcium exchanged with sodium. It is equivalent to gypsum requirement of the soil when 'Ca' is expressed as $CaSO_4$.

Apparatus

- Mechanical shaker
- Burette – 50 ml
- Pipettes – 5ml, 100 ml

Reagents

- Saturated gypsum (calcium sulphate) solution: Add 5 g of chemically pure $CaSO_4.2H_2O$ to one litre of distilled water. Shake vigorously for 10 minutes using a mechanical shaker and filter through Whatman No.1 filter paper.

- 0.01N $CaCl_2$ solution: Dissolve exactly 0.5 g of AR grade $CaCO_3$ powder in about 10 ml of 1:3 diluted HCl. When completely dissolved, transfer to 1 litre volumetric flask and dilute to the mark with distilled water. $CaCl_2$ salt should not be used as it is highly hygroscopic.
- 0.01N Versenate solution: Dissolve 2.0 g of pure EDTA – disodium salt and 0.05g of magnesium chloride (AR grade) in about 50 ml of water and dilute to 1 litre. Titrate a portion of this against 0.01N of $CaCl_2$ solution to standardize.
- Eriochrome BlackT (EBT) indicator: Dissolve 0.5 g of EBT dye and 4.5 g of hydroxylamine hydrochloride in 100 ml of 95% ethanol. Store in a stoppered bottle or flask.
- Ammonium hydroxide-ammonium chloride buffer: Dissolve 67.5 g of pure ammonium chloride in 570 ml of Conc. ammonium hydroxide and dilute to 1 litre. Adjust the pH at 10 using dilute HCl or dilute NH_4OH.

Procedure

1. Weigh 5 g of air-dry soil in 250 ml conical flask.
2. Add 100 ml of the saturated gypsum solution. Firmly put a rubber stopper and shake for 5 minutes on mechanical shaker.
3. Filter the contents through Whatman No.1 filter paper.
4. Transfer 5 ml aliquot of the clear filtrate into a 100 or 150 ml porcelain dish.
5. Add 1 ml of the ammonium hydroxide-ammonium chloride buffer solution and 2to 3 drops of Eriochrome black T indicator.
6. Take 0.01N versenate solution in a 50 ml burette and titrate the contents in the dish until the wine red colour starts changing to sky blue. Volume of versenate used = B.
7. Run a blank using 5 ml of saturated gypsum solution in place of sample aliquot. Volume of versenate solution used = A.

Gypsum requirement (tonnes/ha) = = (A – B) x N x 382

where,

A = ml of EDTA (versenate) used for blank titration

B = ml of ETDA used for soil extract

N = Normality of EDTA solution

7.5 Organic Carbon/Organic Matter

Organic matter estimation in the soil can be done by different methods. Loss of weight on ignition can be used as a direct measure of the organic matter contained in the soil. It can also be expressed as the content of organic carbon in the soil. It is generally assumed that on an average organic matter contains about 58% organic carbon. Organic matter/organic carbon can also be estimated by volumetric and colorimetric methods. However, the use of potassium dichromate ($K_2Cr_2O_7$) involved in these estimations is considered as a limitation because of its hazardous nature. Soil organic matter content can be used as an index of N availability (potential of a soil to supply N to plants) because the content of N in soil organic matter is relatively constant.

7.5.1 Loss of Weight on Ignition

Apparatus

- Sieve
- Beaker
- Oven
- Muffle furnace

Procedure

1. Weigh 5.0 to 10.0 g (to the nearest 0.01 g) sieved (2 mm) soil into an ashing vessel (50 ml beaker or other suitable vessel).
2. Place the ashing vessel with soil into a drying oven set at 105 ^{0}C and dry for 4 hours. Remove the ashing vessel from the drying oven and place in a dry atmosphere. When cooled, weigh to the nearest 0.01 g. Place the ashing vessel with soil into a muffle furnace and bring the temperature to 400 ^{0}C .Ash in the furnace for 4 hours. Remove the ashing vessel from the muffle furnace, cool in a dry atmosphere and weigh to the nearest 0.01 g.

Calculation

Percent organic matter (OM) = $(W1 - W2)x\ 100\ /W1$

Percent organic C = % OM x 0.58

Where,

W_1 is the weight of soil at 105 ^{0}C and W2 is the weight of soil at 400 ^{0}C.

7.5.2 Rapid Titration Method (Walkley and Black, 1934)

Apparatus

- Conical flask - 500 ml
- Pipettes - 2, 10 and 20 ml
- Burette - 50 ml

Reagents

- Phosphoric acid – 85%
- Sodium fluoride solution (chemically pure)
- Sulphuric acid (sp.gr.1.84) containing 1.25% Ag_2SO_4 (silver sulphate)
- Standard IN potassium dichromate: Dissolve 49.04 g of $K_2Cr_2O_7$ in water and dilute to 1litre

- Standard 0.5 N Ferrone ammonium sulphate solution: Dissolve 196g ferrone ammonium sulphatein 800 ml water, add 20 ml concentrated H_2SO_4 and make up the volume to 1 litre
- Diphenylamine indicator: Dissolve 0.5 g reagent grade diphenylamine in a mixture of 20 ml water and 100 ml of concentrated H_2SO_4.

Procedure

1. Take 1.0 g of the prepared soil sample in 500 ml conical flask.
2. Add 10 ml of IN $K_2Cr_2O_7$ solution and 20 ml concentrated H_2SO_4containing Ag_2SO_4.
3. Mix thoroughly and allow the reaction to complete for 30 minutes.
4. Dilute the reaction mixture with 200 ml water and add 10 ml H_3PO_4.
5. Add 0.5 g of NaF solution and 1ml of diphenylamine indicator.
6. Titrate the solution with standard 0.5 N $FeSO_4$ solution till the colour change from blue-violet to a brilliant green colour.
7. Simultaneously a blank is sun without soil.

Calculation

$$\text{Percent organic Carbon (X)} = \frac{10(S-T)\times 0.003}{S}\times\frac{100}{\text{wt.of soil}}$$

Since one gram of soil is used, this equation simplifies to:

$$\frac{3(S-T)}{S}$$

Where,

S = ml $FeSO_4$ solution required for blank

T = ml $FeSO_4$ solution required for soil sample

3 = Eq W of C (weight of C is 12, valency is 4, hence Eq W is 12÷4 = 3.0)

0.003 = weight of C (1000 ml IN $K_2Cr_2O_7$ = 3 g C. Thus, 1 ml IN ss$K_2Cr_2O_7$ = 0.003 g C) Organic Carbon recovery is estimated to be about 77%. Therefore, actual amount oforganic carbon (Y) will be:

Percent value of organic carbon obtained x 100/77

Or Percentage value of organic carbon x 1.3

Percent Organic matter = Y x 1.724 (organic matter contains 58 % organic carbon, hence 100/58 = 1.724)

Note: Published organic C to total organic matter conversion factor for surface soils vary from 1.724 to 2.0. A value of 1.724 is commonly used, although whenever possible the appropriate factor be determined experimentally for each type of soil.

7.5.3 Colorimetric Method

Apparatus

- Spectrophotometer
- Conical flask -100 ml
- Pipettes - 2 ,5 and 10 ml

Reagents

- Standard (AR grade) potassium dichromate
- Concentrated sulphuric acid containing 1.25% Ag_2SO_4
- Sucrose (AR quality)

Procedure

1. Preparation of standard curve: Sucrose is used as a primary standard as carbon source. Take different quantities of sucrose (1 mg to 20 mg) in 100 ml flasks. Add 10 ml standard 1 N $K_2Cr_2O_7$ and 20 ml of concentrated H_2SO_4 in each flask. Swirl the flasks and leave for 30 minutes. A blank is also prepared in the similar way without adding sucrose. Green colour develops which is read on spectrophotometer at 660 nm, after adjusting the blank to zero. The reading so obtained is plotted against mg of sucrose as carbon source (carbon = wt. of sucrose x 0.42 because carbon content of sucrose is 42%) or against mg C directly.
2. Take 1 g of soil in 100 ml conical flask.
3. Add 10 ml of 1N $K_2Cr_2O_7$ and 20 ml of conc. H_2SO_4 containing 1.25percent of Ag_2SO_4.
4. Stir the reaction m ixture and allow to stand for 30 minutes.
5. The green colour of chromium sulphate so developed is read on a spectrophotometer at 660 nm (red) filter after setting the blank, prepared in the similar manner, at zero.

Calculation

The carbon content of the sample is found out from the standard curve which shows the carbon content (mg of carbon v/s spectrophotometer readings as absorbance).

Percent C = mg C observed x 100/1000 (observed reading is for 1 g soil, expressed as mg).

Percent OM = % C x 1.724

7.6 Total Nitrogen (Kjeldahl Method)

Total N includes all forms of inorganic N, like NH_4 –N, NO_3 -N and also NH_2(Urea) –N, and the organic N compounds like proteins, amino acids and other derivatives. Depending upon the form of N present in a particular sample, specific method is to be adopted for getting the total nitrogen value. While the organic N materials can be converted into simple inorganic ammoniacal salt by digestion with sulphuric acid, for reducing nitrates into ammoniacal form, use of salicylic acid or Devarda's alloy is made in the modified Kjeldahl method. At the end of digestion, all organic and inorganic salts are converted into ammonium form which is distilled and estimated by using standard acid. As the precision of the method depends upon complete conversion of organic N into NH_4 - N, the digestion temperature and time, solid: acid ratio and the type of catalyst used have an important bearing on the method. The ideal temperature for digestion is 320 °C – 370 °C. At lower temperature, the digestion may not be complete, while above 410 °C, the loss of NH_3 may occur. The salt:acid (weight:volume) ratio should not be less than 1:1 at the end of digestion. Commonly used catalysts to hasten the digestion process are $CuSO_4$ or Hg. Potassium sulphate is added to raise the boiling point of the acid so that loss of acid by volatilization is prevented.

Apparatus

- Kjeldahl digestion and distillation unit
- Conical flasks
- Burettes
- Pipettes

Reagents

- Sulphuric acid – H_2SO_4 (93-98%)
- Copper sulphate – $CuSO_{4.}H_2O$ (AR grade)
- Potassium sulphate or anhydrous sodium sulphate (AR grade)
- 35% sodium hydroxide solution: Dissolve 350 g solid NaOH in water and dilute to one litre
- 0.1M NaOH: Prepare 0.1M NaOH by dissolving 4.0 g NaOH in water and make volume to 1 litre. Standardize against 0.1N potassium hydrogen phthalate o rstandard H_2SO_4
- 0.1M HCl or 0.1M H_2SO_4: Prepare approximately 0.1M acid solution and standardize against 0.1M sodium carbonate
- Methyl red indicator
- Salicyclic acid for reducing NO_3 to NH_4, if present in the sample
- Devarda's alloy for reducing NO_3 to NH_4, if present in the sample.

Procedure

1. Weigh 1 g sample of soil. Place in Kjeldahl flask .
2. Add 0.7 g copper sulphate, 1.5 g K_2SO_4 and 30 ml H_2SO_4.
3. Heat gently until frothing ceases. If necessary, add small amount of paraffin or glass beads to reduce frothing.

4. Boil briskly until solution is clear and then continue digestion for at least 30 minutes. Remove the flask from the heater and cool, add 50 ml water and transfer to distilling flask.
6. Take accurately 20–25 ml standard acid (0.1M HCl or 0.1M H_2SO_4) in the receiving conical flask so that there will be an excess of at least 5 ml of the acid.

   Add 2-3 drops of methyl red indicator. Add enough water to cover the end of the condenser outlet tubes.
7. Add 30 ml of 35% NaOH in the distilling flask in such a way that the contents do not mix.
8. Heat the contents to distil the ammonia for about 30-40 minutes.
9. Remove receiving flask and rinse outlet tube into receiving flask with a small amount of distilled water.
10. Titrate excess acid in the distillate with 0.1M NaOH.
11. Determine blank on reagents using same quantity of standard acid in a receiving conical flask.

$$\text{Percent N} = \frac{1.401(V_1 M_1 - V_2M_2) - (V_3M_1 - V_4M_2)}{W} \times df$$

Where, W

V_1 - ml of standard acid taken in receiving flask for samples

V_2 - ml of standard NaOH used in titration

V_3 - ml of standard acid taken to receiving flask for blank

V_4 - ml of standard NaOH used in titrating blank

M_1 - Molarity of standard acid

M_2 - Molarity of standard NaOH

W - Weight of sample taken (1 g)

df - Dilution factor of sample (if 1 g was taken for estimation, the dilution factor will be 100).

Note: 1000 ml of 0.1 M HCl or 0.1 M H_2SO_4 = 1.401 g Nitrogen

Precautions

The material after digestion should not solidify.

No NH_4 should be lost during distillation.

If the indicator changes colour during distillation, determination must be repeated using either a smaller sample weight or a larger volume of standard acid.

7.6.1 Mineralizable Nitrogen by Alkaline Permanganate Method (Subbia and Asija, 1956)

In case of soils, mineralizable N (also organic C) is estimated as an index of available nitrogen content and not the total nitrogen content. The easily mineralizable nitrogen is estimated using alkaline $KMnO_4$, which oxidizes and hydrolyses the organic matter present in the soil. The liberated ammonia is condensed and absorbed in boric acid, which is titrated against standard acid. The method has been widely adopted to get a reliable index of nitrogen availability in soil due to its rapidity and reproducibility. The process of oxidative hydrolysis is, however, a progressive one and thus, a uniform time and heating temperature should be allowed for best results. Use of glass beads checks bumping while liquid paraffin checks frothing during heating as is recommended in total N estimation by Kjeldahl method.

Apparatus

- . Nitrogen distillation unit, preferably with six regulating heating elements.
- Conical flasks, pipettes, burette, etc.

Reagents

- 0.32% $KMnO_4$: Dissolve 3.2 g of $KMnO_4$ in distilled water and make the volume to one litre.
- 2.5% NaOH: Dissolve 25 g of sodium hydroxide pellets in water and make the volume to one litre.
- 2% Boric acid: Dissolve 20 g of boric acid powder in warm water by stirring and dilute to one litre.
- Mixed Indicator: Dissolve 0.066 g of methyl red and 0.099 g of bromocresol green in 100 ml of ethyl alcohol. Add 20 ml of this mixed indicator to each litre of 2% boric acid solution.
- 0.1M Potassium Hydrogen Phthalate: Dissolve 20.422 g of the salt in water and dilute to one litre. This is a primary standard and does not require standardization.
- 0.02N H_2SO_4: take 5.6 ml of conc. H_2SO_4 to about 1 litre of distilled water. This will give 0.2 N H_2SO_4. Now take 100 ml. of this 0.2 N H_2SO_4& make it to 1 litre with distilled waterto get 0.02 N H_2SO_4. Standardize it against 0.1 M NaOH solution.
- 0.1M NaOH: Dissolve 4g NaOH in 100 ml distilled water. Standardize against potassium hydrogen phthalate.

Procedure

1. Weigh 20 g of soil sample in an 800 ml Kjeldahl flask.
2. Moisten the soil with about 10 ml of distilled water, wash down the soil, if any, adhering to the neck of the flask.
3. Add 100 ml of 0.32% of $KMnO_4$ solution.

4. Add a few glass beads or broken pieces of glass rod.
5. Add 2-3 ml of paraffin liquid, avoiding contact with upper part of the neck of the flask to prevent frothing & bumping respectively during distillation.
6. Measure 20 ml of 2% boric acid containing mixed indicator in a 250 ml conical flask and place it under the receiver tube. Dip the receiver tube in the boric acid.
7. Run tap water through the condenser.
8. Add 100 ml of 2.5% NaOH solution and immediately attach to the rubber stopper fitted in the alkali trap.
9. Switch the heaters on and continue distillation until about 100 ml of distillate is collected.
10. First remove the conical flask containing distillate and then switch of the heater to avoid back suction.
11. Titrate the distillate against 0.02M H_2SO_4 taken in burette until pink colour starts appearing.
12. Run a blank without soil.
13. Carefully remove the Kjeldahl flask after cooling and drain the contents in the sink.

Calculation

Volume of acid used to neutralize ammonia in the sample = A – B ml N content in the test sample = (A – B) x 0.56 mg

Percent Nitrogen (A – B) x 0.56 x 5

Where,

A = Volume of 0.02N H_2SO_4 used in titration against ammonia absorbed in boric acid.

B = Volume of 0.02N sulphuric acid used in blank titration.

1 ml of 0.02N sulphuric acid = 0.56 mg N (1 000 ml of 1N H_2SO_4 = 14 g Nitrogen).

Wt. of soil sample = 20 g. Thus, factor for converting into % Nitrogen

= 100/20 = **5**

Precautions

- Check all the joints of the Kjeldahl apparatus to prevent any leakage and loss of ammonia.
- Hot Kjeldahl flasks should neither be washed immediately with cold water nor allowed to cool for long to avoid deposits to settle at the bottom which are difficult to remove.
- In case frothing takes place and passes through to the boric acid, such samples should be discarded and fresh distillation done.

- Opening ammonia bottles in the laboratory should be strictly prohibited while distillation is on. The titration should be carried out in ammonia free atmosphere.
- In case the titration is not to be carried out immediately, the distillate should bestored in ammonia free cupboards after tightly stoppering the flasks.

7.6.2 Inorganic Nitrogen - NO_3^- and NH_4^+

Inorganic N in soil is present predominantly as NO_3^- and NH_4^+. Nitrite is seldom present in detectable amount, and its determination is normally unwarranted except in neutral to alkaline soils following the application of NH_4 or NH_4-forming fertilizers.

Nitrate is highly soluble in water, and a number of solutions including water have been used as extractants. These include saturated 0.35% $CaSO_4$ $2H_2O$ solution, 0.03M NH_4F, 0.015M H_2SO_4, 0.01M $CaCl_2$, 0.5M $NaHCO_3$ (pH 8.5), 0.01M $CuSO_4$, 0.01M $CuSO_4$ containing Ag_2SO_4 and 2.0M KCl. Exchangeable NH_4 is defined as NH_4 that can be extracted at room temperature with a neutral K salt solution. Various molarities have been used, such as 0.05M K_2SO_4, 0.1M KCl, 1.0M KCl, and 2.0M KCl. The potential of a soil to mineralize N as measured by N availability indexes (OM, OC and even total N) is fairly constant from year to year, making it unnecessary to make that type of determination each year. However, it is still necessary to take into consideration the initial amount of available N (inorganic N: NO_3 – N and/or NH_4 – N) in the rooting zone at or near planting time for better prediction of N fertilizer need. In contrast to N index tests, this type tests must be made each year, especially when there is possibility of residual inorganic N remaining from previous application or fallow period. The methods for the determination of NO_3- N and NH_4- N are even more diverse than the methods of extraction (Keeney and Nelson 1982). These range from specific ion electrode to colorimetric techniques, microdiffusion, steam distillation, and flow injection analysis. Steam distillation is still a preferred method when using 15N. However, for routine analysis phenoldisulfonic acid method for NO_3 and indophenol blue method for NH_4 estimation have been described

7.6.3 Nitrate by Phenoldisulfonic Acid Method

One of the major difficulties in estimating NO_3 in soils by colorimetric methods is obtaining a clear colourless extract with low contents of organic and in organic substances which interfere with the colorimetric method. In arid and salt affected soils, chloride (Cl) is the major anion which interferes with colour development of the phenoldisulfonic acid method. Therefore, if chloride concentration is more than 15μg/g soil, it should be removed before analysis by the use of Ag_2SO_4 to precipitate chloride as AgCl. The Ag_2SO_4 is added to the extract or to the reagent used for extraction, and the AgCl is removed by filtration or centrifugation after precipitation of the excess Ag_2SO_4 by basic reagent such as $Ca(OH)_2$ or $MgCO_3$. It is necessary to remove the excess Ag^+ before analysis of the extract because it also interferes with the phenoldisulfonic acid method of determining NO_3.

Apparatus

- Reciprocating shaker
- Heavy-duty hot plate
- Spectrophotometer
- Dispenser
- Erlenmeyer flask
- Beakers
- Glass rod

Reagents

- Phenoldisulfonic acid (phenol 2,4-disulfonic acid): Transfer 70 ml pure liquid phenol (carbolic acid) to an 800 ml Kjeldahl flask. Add 450 ml concentratedH_2SO_4 while shaking. Add 225 ml fuming H_2SO_4 (13-15% SO_3). Mix well. Place Kjeldahl flask (loosely stoppered) in a beaker containing boiling water and heatfor 2 hours. Store resulting phenoldisulfonic acid $[C_6H_3OH(HSO_3)_2]$ solution in a glass-stoppered bottle.
- Dilute ammonium hydroxide solution (about 7.5M NH_4OH): Mix one part NH_4OH (specific gravity 0.90) with one part H_2O.
- Copper sulfate solution (0.5M): Dissolve 125 g $CuSO_4.5H_2O$ in 1 litre of distilled water.
- Silver sulfate solution (0.6%): Dissolve 6.0 g Ag_2SO_4 in 1 litre of distilled water. Heat or shake well until all salt is dissolved.
- Nitrate-extracting solution: Mix 200 ml of 0.5M copper sulfate solution and 1 litre 0.6% silver sulfate solution and dilute to 10 litres with water. Mix well.
- Standard nitrate solution (100 µg NO_3-N/ml, stock solution): Dissolve 0.7221 g KNO_3 (oven dried at 105 °C) in water and dilute to 1 litre. Mix thoroughly.
- Standard nitrate solution (10 µg NO_3-N/ml; working solution): Dilute 100 ml of100 µgNO_3-N/ml stock solution to 1 litre with water. Mix well.
- Calcium hydroxide, AR grade powder (free of NO_3).
- Magnesium carbonate, AR grade powder (free of NO_3).

Procedure

1. Place about 5 g soil in an Erlenmeyer flask.
2. Add 25 ml of nitrate-extracting solution.
3. Shake contents for 10 minutes.
4. Add about 0.2 g $Ca(OH)_2$ and shake for 5 minutes.
5. Add about 0.5 g $MgCO_3$ and shake for 10-15 minutes.

6. Allow to settle for a few minutes.
7. Filter through a Whatman filter paper No. 42.
8. Pipette 10 ml of clear filtrate into a 100 ml beaker. Evaporate to dryness on a hotplate at low heat in a fume hood (free of HNO_3 fumes). Do not continue heating beyond dryness.
9. When completely dry, cool residue, add 2 ml phenoldisulfonic acid rapidly (from a burette having the tip cut off) covering the residue quickly. Rotate the beaker so that reagent comes in contact with all residual salt. Allow to stand for 10-15minutes.
10. Add 16.5 ml cold water. Rotate the beaker to dissolve residue (or stir with a glass rod until all residue is in solution).
11. After the beaker gets cool, add 15 ml dilute NH_4OH slowly until the solution is distinctly alkaline as indicated by the development of a stable yellow color.
12. Add 16.5 ml water (volume becomes 50 ml). Mix thoroughly.
13. Read concentration of NO_3-N at 415 nm, using the standard curve.
14. Preparation of standard curve: Take 0, 2, 5, 8, and 10 ml of the 10 μg NO_3/ ml working solution in respective 100 ml beakers, add 10 ml NO3-extracting solution and evaporate to dryness. Follow steps 9 to 13, using these standard solution shaving 0, 0.4, 1.0, 1.6 and 2.0 μg NO_3-N/ml. Prepare a standard curve to be used for estimation of NO_3 in the sample.

Calculation

NO_3^- in test solution (μg/ml) =

$$\frac{\text{Vol. after colour development(ml)}}{\text{Vol/ evaporated(ml)}} \times \frac{\text{Vol. of extracting soln. (ml)}}{\text{Wt of oven-dried soil(g)}}$$

7.6.4 Ammonium by Indophenol Blue Method

The phenol reacts with NH_4 in the presence of an oxidizing agent such as hypochlorite to form a coloured complex in alkaline condition. The addition of sodium nitroprusside as a catalyst in the reaction between phenol and NH_4 increases the sensitivity of the method by many folds. The addition of EDTA is necessary to complex divalent and trivalent cations present in the extract. Otherwise, it forms precipitate at the pH of 11.4 – 12 used for color development, and this turbidity would interfere with formation of the phenol–NH_4 complex. The first step is the extraction of exchangeable Ammonium.

Apparatus

- Erlenmeyer flask
- Volumetric flask
- Shaker
- Spectrophotometer

Reagents

- Potassium chloride (KCl) solution, 2M: Dissolve 150 g AR grade KCl in 1 litre distilled water.
- Standard ammonium (NH_4^+) solution: Dissolve 0.4717 g of ammonium sulfate $(NH_4)_2SO_4$ in water, and dilute to a volume of 1 litre. If pure dry $(NH_4)_2SO_4$ isused, the solution contains 100 µg of NH_4-N/ml. Store the solution in a refrigerator. Immediately before use, dilute 4 ml of this stock NH_4^+ solution to200 ml. The resulting working solution contains 2 µg of NH_4-N/ml. Accordingly, various concentrations of standard solution to be made for the standard curve.
- Phenol-nitroprusside reagent: Dissolve 7 g of phenol and 34 mg of sodium nitroprusside [disodium pentacyanonitrosylferrate, $Na_2Fe(CN)_5NO.2H_2O$] in 80ml of NH_4^+-free water and dilute to 100 ml. Mix well, and store in a dark-coloredbottle in the refrigerator.
- Buffered hypochlorite reagent: Dissolve 1.480 g of sodium hydroxide (NaOH) in70 ml of NH_4^+-free water, add 4.98 g of sodium monohydrogen phosphate(Na_2HPO_4) and 20 ml of sodium hypochlorite (NaOCl) solution (5-5.25%NaOCl). Use less or more hypochlorite solution if the concentration is higher or lower, respectively than that is indicated above. Check the pH to ensure a value between 11.4 and 12.2. Add a small amount of additional NaOH if required to raise the pH. Dilute to a final volume of 100 ml.
- Ethylene diaminetetraacetic acid (EDTA) reagent: Dissolve 6 g of ethylene diamine tetraacetic acid disodium salt (EDTA disodium) in 80 ml of deionized water, adjust to pH 7, mix well, and dilute to a final volume of 100 ml.

Extraction

1. Place 10 g of soil in a 250 ml wide-mouth Erlenmeyer flask and add 100 ml of 2M KCl.
2. Put stopper and shake the flask on a mechanical shaker for 1 hour.
3. Allow the soil-KCl suspension to settle (about 30 min) till the supernatant is clear.
4. If the KCl extract can not be analyzed within 24 hours, then filter the soil-KCl suspension (Whatman no. 42 filter paper) and store in the refrigerator.Aliquots from this extract is used for the NH_4 estimation.

Estimation

1. Pipette an aliquot (not more than 5 ml) of the filtered 2M KCl extract containing between 0.5 and 12µg of NH_4-N into a 25 ml volumetric flask.
2. Add 1 ml of the EDTA reagent, and mix the content of the flask.
3. Allow the content to stand for 1 minute, then add 2 ml of the phenol-nitroprusside reagent, followed by 4 ml of the buffered hypochlorite reagent, and immediately dilute the flask to volume (25 ml) with NH_4^+ - free water and mix well.

4. Place the flask in a water bath maintained at 40° C, and allow it to remain for 30min.
5. Remove the flask from the bath, cool to room temperature, and determine the absorbance of the coloured complex at a wavelength of 636 nm against a reagent blank solution.
6. Determine the NH_4-N concentration of the sample by reference to a calibration curve plotted from the results obtained with 25 ml standard samples containing 0,2, 4, 6, 8, 10, and 12µg of NH_4-N/ml.
7. To prepare this curve, add an appropriate amount of 2M KCl solution (same volume as that used for aliquots of soil extract, i.e. about 5 ml)) to a series of 25-ml volumetric flasks. Add 0, 1, 2, 3, 4, 5, and 6 ml of the 2 µg NH_4^{-N}/ml solution to the flasks, and measure the intensity of blue colour developed with these standards by the procedure described above for the analysis of unknown extracts.

Calculation

NH_4-N in the sample as noted from the standard curve= A (µg/ml)

$$\text{µg of NH N in 1 g soil} = \frac{A \times 100(\text{total vol. of extract})}{5(\text{vol. of extract estimted})} \times \frac{1}{10(\text{wt. of soil})}$$

where,

Weight of the soil taken for estimation = 10 g

Total volume of extract = 100 ml

Volume of extract taken for estimation = 5 ml

7.7 Available Phosphorus

Two methods are most commonly used for determination of available phosphorus in soils: Bray's Method No.1 for acidic soils and Olsen's Method for neutral and alkaline soils. In these methods, specific coloured compounds are formed with the addition of appropriate reagents in the solution, the intensity of which is proportionate to the concentration of the element being estimated. The colour intensity is measured spectrophotometrically. In spectrophotometric analysis, light of definite wavelength (not exceeding say 0.1 to 1.0 nm in band width) extending to the ultraviolet region ofthe spectrum constitutes the light source. The photoelectric cells in spectrophotometer measure the light transmitted by the solution. A spectrophotometer, as its name implies, is really two instruments in one cabinet – a spectrometer and a photometer. A spectrometer is a device for producing coloured light of any selected colour (or wavelength) and, when employed as part of aspectrophotometer, is usually termed as monochromator and is generally calibrated inwavelengths (nm). A photometer is a device for measuring the intensity of the light, and when incorporated in a spectrophotometer is used to measure the intensity of the monochromatic beam produced by the associated monochromator. Generally, the photometric measurement is made first with a reference liquid and then with acoloured sample contained in similar cells interposed in the light beam: the ratio of the two intensity measurements being a measure of the opacity of the sample at the wavelength of the test.

White light covers the entire visible spectrum 400-760 nm

7.7.1 Bray's Method No. 1 (Bray and Kurtz, 1945) for Acid Soils

Apparatus

- Spectrophotometer
- Pipette - 2 ml, 5 ml, 10 ml and 20 ml
- Bearkers/flasks - 25 ml, 50 ml, 100 ml and 500 ml

Reagents

- Bray & Kurtz extracting solution (0.03N NH_4F in 0.025N HCl): Dissolve 11.1 g of AR grade NH_4F in 100 ml of distilled water, filter, and add to the filtrate 1 litre of water containing 20 ml of concentrated HCl, make up the volume to 10 litres with distilled water. It can be stored (one year).
- 1.5 % Ammonium Molybdate reagent: Dissolve 15g $(NH_4)_6$ MO_7 $O_{24}.4H_2O$ in 300 ml warm distilled water. Add the solution to 342 ml of concentrated HCl solution gradually with stirring. Dilute to 1litre with distilled water.
- 40% Stannous chloride solution (Stock Solution): Dissolve 10 g SnCl. $2H_2O$ in 25 ml of concentrated HCl. Store the solution in a glass stoppered bottle.
- Stannous chloride solution (Working Solution): Dilute 0.5 ml of the stock solution of stannous chloride to 66.0 ml with distilled water just before use. Prepare fresh dilute solution every working day.

Procedure

1. Preparation of the Standard Curve: Dissolve 0.1916 g of pure dry KH_2PO_4 in 1litre of distilled water. This solution contains 0.10 mg P_2O_5/ml. Preserve this as a stock standard solution of phosphate. Take 10 ml of this solution and dilute it to 1litre with distilled water. This solution contains 1 μg P_2O_5/ ml (0.001 mg P_2O_5/ml).Take 1, 2, 4, 6 and 10 ml of this solution in separate 25 ml flasks. Add to each, 5ml of the extractant solution, 5 ml of the molybdate reagent and dilute with distilled water to about 20 ml. Add 1 ml dilute $SnCl_2$ solution, shake again and dilute to the 25 ml mark. After 10 minutes, read the blue colour of the solution on the spectrophotometer at 660 nm wavelength. Plot the absorbance reading against μg P_2O_5 and join the points.
2. Extraction: Add 50 ml of the Bray's extractant No. 1 to the 100 ml conical flask containing 5 g soil sample. Shake for 5 minutes and filter.
3. Development of colour: Take 5 ml of the filtered soil extract with a bulb pipette in a 25 ml measuring flask; deliver 5 ml of the molybdate reagent with an automatic pipette, dilute to about 20 ml with distilled water, shake and add 1 ml of the dilute $SnCl_2$ solution with a bulb pipette. Fill to the 25

ml mark and shake thoroughly. Read the blue colour after 10 minutes on the spectrophotometer at 660 nm wavelength after setting the instrument to zero with the blank prepared similarly but without the soil.

Calculation

$$P(kg / ha) = \frac{A}{10000000} \times \frac{50}{5} \times \frac{20000000}{5} = 4A$$

Where,

Weight of the soil taken = 5 g

Volume of the extract = 50 ml

Volume of the extract taken for estimation = 5 ml

Volume made for estimation (dilution = 5 times) = 25 ml

Amount of P observed in the sample on the standard curve = A (μg).

Wt. of 1 ha of soil upto a depth of 22 cm is taken as 2 million kg.

7.7.2 Sodium Bicarbonate Extractable P or Olsen's Method (Olsen, et al, 1954)

Apparatus

Same as for Bray's Method No. 1.

Reagents

- Bicarbonate extractant (0.5 M Na H CO_3): Dissolve 42 g Sodium bicarbonate in 1 litre of distilled water and adjust the pH to 8.5 by addition of dilute NaOH or HCl. Filter, if necessary.
- Activated carbon – Darco G 60.
- Molybdate reagent: Same as for the Bray's Method.
- Stannous chloride solution: Same as in Bray's Method.

Procedure

1. Preparation of the standard curve: Procedure is the same as in Bray's Method .
2. Extraction: Add 50 ml of the bicarbonate extractant to 100 ml conical flask, containing 2.5 g soil sample. Add 1 g activated carbon. Shake for 30 minutes onthe mechanical shaker and filter.
3. Development of Colour: Procedure same as described under the Bray's Method.

Calculation

Same as described under the Bray's Method.

Caution

In spite of all precautions, intensity of blue colour changes slightly with every batch of molybdate reagent. It is imperative to check standard curve every day by using 2 or 3 dilutions of the standard phosphate solution. If the standard curve does not tally, draw a new standard curve with fresh molybdate reagent.

7.8 Available Potassium

Flame Photometeric Method (Toth and Prince, 1949)

Potassium present in the soil is extracted with neutral normal ammonium acetate solution. This is considered as plant available K in the soils. It is estimated with the help of flame photometer. This is a well accepted method. Soil is shaken with a neutral normal solution of ammonium acetate. During extraction ammonium ions replace potassium ions absorbed on the soil colloids.

Apparatus

- Multiple Dispenser or automatic pipette – 25 ml
- Flasks and beakers - 100 ml
- Flame Photometer

Reagents

- Molar neutral ammonium acetate solution: Dissolve 77 g of ammonium acetate ($NH_4C_2H_3O_2$) in 1 litre of water. Check the pH with bromothymol blue or with a pH meter. If not neutral, add either ammonium hydroxide or acetic acid as per the need to neutralize it to pH 7.0.
- Standard potassium solution: Dissolve 1.908 g pure KCl in 1 litre of distilled water. This solution contains 1 mg K/ml. Take 100 ml of this solution and dilute to 1 litre with ammonium acetate solution. This gives 0.1 mg K/ml as stock solution.
- Working potassium standard solutions: Take 0, 5, 10, 15 and 20 ml of the stock solution separately and dilute each to 100 ml with the N ammonium acetate solution. These solutions contain 0, 5, 10, 15 and 20 μg K/ml, respectively.

Procedure

1. Preparation of the Standard Curve: Set up the flame photometer by atomizing 0 and 20 μg K/ml solutions alternatively to 0 and 100 reading. Atomize intermediate working standard solutions and record the readings. Plot these readings against the respective potassium contents and connect the points with a straight line to obtain a standard curve.
2. Extraction: Add 25 ml of the ammonium acetate extractant to conical flask fixed in a wooden rack containing 5 g soil sample. Shake for 5 minutes and filter.
3. Determine potash in the filtrate with the flame photometer.

Calculation

$$K(kg / ha) = A \times 25 / 5 \times \frac{20000000}{1000000} = 10A$$

Where,

A = content of K (μg) in the sample, as read from the standard curve:

Weight of 1 ha of soil upto a plough depth of 22 cm is approx. 2 million kg.

7.9 Available Sulphur

Available sulphur in mineral soils occurs mainly as adsorbed SO_4 ions. Phosphate ions (as monacalcium phosphate) are generally preferred for replacement of the adsorbed SO_4 ions. The extraction is also carried out using $CaCl_2$ solution. However, the former is considered to be better for more efficient replacement of SO_4 ions. Use of Ca salts have a distinct advantage over those of Na or K as Ca prevents deflocculation in heavy textured soils and leads to easy filtration. SO_4 in the extract can be estimated turbidimetrically using a spectrophotometer. A major problem arises when the amount of extracted sulphur is too low to be measured. To overcome this problem, seed solution of known S concentration is added to the extract to raise the concentration to easily detectable level. Barium sulphate precipitation method is described here.

Apparatus

- Spectrophotometer
- Mechanical shaker
- Volumetric flask

Reagents

- Mono-calcium phosphate extracting solution (500 mg P/litre): Dissolve 2.035 g of $Ca(H_2PO_4)_2.H_2O$ in 1 litre of water.
- Gum acacia-acetic acid solution: Dissolve 5g of chemically pure gum acacia powder in 500 ml of hot water and filter in hot condition through Whatman No.42 filter paper. Cool and dilute to one litre with dilute acetic acid.
- Barium chloride: Pass AR grade $BaCl_2$ salt through 1 mm sieve and store for use.
- Standard stock solution (2000 mg S/litre): Dissolve 10.89g of oven-dried AR grade potassium sulphate in 1 litre water.
- Standard working solution (10 mg S/litre): Measure exactly 2.5 ml of the stock solution and dilute to 500 ml.
- Barium sulphate seed suspension: Dissolve 18 g of AR grade $BaCl_2$ in 44 ml of hot water and add 0.5 ml of the standard stock solution. Heat the content to boiling and then cool quickly. Add 4 ml of gum acacia-acetic acid solution to it. Prepare a fresh seed suspension for estimation everyday.

- Dilute nitric acid (approx 25%): Dilute 250 ml of AR grade conc. HNO_3 to one litre.
- Acetic-phosphoric acid: Mix 900 ml of AR grade glacial acetic acid with 300 ml of H_3PO_4 (AR grade).

Procedure

1. Weigh 20 g of soil sample in a 250 ml conical flask. Add 100 ml of the monocalcium phosphate extracting solution (500 mg P/litre) and shake for one hour. Filter through Whatman No.42 filter paper.
2. Take 10 ml of the clear filtrate into a 25 ml volumetric flask.
3. Add 2.5 ml of 25% HNO_3 and 2 ml of acetic-phosphoric acid. Dilute to about 22ml, stopper the flask and shake well, if required.
4. Shake the $BaSO_4$ seed suspension and then add 0.5 ml of it, and 0.2 g of $BaCl_2$ crystals. Stopper the flask and invert three times and keep.
5. After 10 minutes, invert 10 times. Again after 5 minutes invert for 5 times.
6. Allow to stand for 15 minutes and then add 1 ml of gum acacia-acetic acid solution.
7. Make up the volume to 25 ml, invert 3 times and keep aside for 90 minutes.
8. Invert 10 times and measure the turbidity intensity at 440 nm (blue filter).
9. Run a blank side by side.
10. Preparation of standard curve:

- Put 2.5, 5.0, 7.5, 10.0, 12.5, 15.0 ml of the working standard solution (10 mg S/litre) into a series of 25 ml volumetric flasks to obtain 25, 50, 75, 100, 125 and 150 mg S.
- Proceed to develop turbidity as described above for sample aliquots.
- Read the turbidity intensity and prepare the curve by plotting readings against sulphur concentrations (in µg in the final volume of 25 ml).

Calculation

Available Sulphur (SO_4^{2-} -S)) in soil (mg/kg) = W X100 /10 x20 =W/2

Where,

W stands for the quantity of sulphur in "mg" as obtained on X-axis against an absorbance reading (Y-axis) on standard curve "20" is the weight of the soil sample in g, "100" is the volume of the extractant in ml "10" is the volume of extractant in "ml" in which turbidity is developed.

7.10 Determination of Exchangeable Calcium and Magnesium

Exchangeable cations are usually determined in a neutral normal ammonium acetate extract of soil. Extraction is carried out by shaking the soil: extractant mixture followed by filtration or centifugation. Calcium and magnesium are determined either by EDTA titration method or by atomic absorption spectrophotometer after the removal of ammonium acetate and organic matter.

It may be noted that in soils appreciable amount of soluble calcium and magnesium may be present. Hence, these water soluble cations are estimated in the 1:2 soil water extract and deducted from ammonium acetate extractable calcium and magnesium (since ammonium acetate also extracts water soluble cations) to obtain exchangeable calcium and magnesium. To obtain soil water extract, generally 25g soil and 50ml of water suspension is shaken for 30 minutes on a mechanical shaker and filtered. The method of estimation in the water extract (water soluble cations) and ammonium acetate extract (exchangeable cation) is same. The EDTA titration method developed by Chang and Bray (1951) is preferred on account of its accuracy, simplicity and speed. The method is based on the principle that calcium, magnesium and a number of other cations form stable complexes with versenate (ethylene diamine tetra acetic acid disodium salt) at different pH. The interference of Cu, Zn, Fe, Mn is prevented by the use of 2% NaCN solution or carbamate. Usually in irrigation waters and water extracts of soil, the quantities of interfering ions are negligible and can be neglected. A known volume of standard calcium solution is titrated with standard versenate 0.01N solution using muroxide (ammonium purpurate) indicator in the presence of NaOH solution. The end point is a change of colour from orange red to purple at pH 12 when the whole of calcium forms a complex with EDTA.

7.10.1 Calcium by Versenate (EDTA) Method

Apparatus

- Shaker
- Porcelain dish
- Beakers
- Volumetric/conical flask.

Reagents

- Ammonium chloride – ammonium hydroxide buffer solution: Dissolve 67.5 g ammonium chloride in 570 ml of conc. ammonium hydroxide and make to 1 litre.
- Standard 0.01N calcium solution: Take accurately 0.5 g of pure calcium carbonate and dissolve it in 10 ml of 3N HCl. Boil to expel CO_2 and then make the volume to 1 litre with distilled water.
- EDTA solution (0.01N): Take 2.0 g of versenate, dissolve in distilled water and make the volume to 1 litre. Titrate it with 0.01N calcium solution and make necessary dilution so that its normality is exactly equal to 0.01N.
- Muroxide indicator powder: Take 0.2 g of muroxide (also known as ammonium purpurate) and mix it with 40 g of powdered potassium sulphate. This indicator should not be stored in the form of solution, otherwise it gets oxidized.

- Sodium diethyl dithiocarbamate crystals: It is used to remove the interference of other metal ions.
- Sodium hydroxide 4N: Prepare 16% soda solution by dissolving 160 g of pure sodium hydroxide in water and make the volume to 1 litre. This will give pH 12.

Procedure

1. Take 5 g air dried soil sample in 150 ml conical flask and add 25 ml of neutral normal ammonium acetate. Shake on mechanical shaker for 5 minutes and filterthrough Whatman filter paper No.1.
2. Take a suitable aliquot (5 or 10 ml) and add 2-3 crystals of carbamate and 5 ml of 16% NaOH solution.
3. Add 40-50 mg of the indicator powder. Titrate it with 0.01N EDTA solution till the colour gradually changes from orange red to reddish violet (purple). It is advised to add a drop of EDTA solution at an interval of 5 to 10 seconds, as the change of colour is not instantaneous.
4. The end point must be compared with a blank reading. If the solution is over titrated, it should be back titrated with standard calcium solution and exact volume used is thus found.
5. Note the volume of EDTA used for titration.

Calculation

If N_1 is normality of Ca^{++} and V_1 is volume of aliquot taken and N_2V_2 are the normality and volume of EDTA used, respectively, then,$N_1V_1 = N_2V_2$

$$\text{Or } N_1 = N_2V_2/V_1 = \frac{\text{Normality of EDTA} \times \text{Vol. of EDTA}}{\text{ml of aliquot taken}}$$

When expressed on soil weight basis,

$$Ca^{2+} \text{ meq/100 g soil} = \frac{100}{\text{Wt. of soil}} \times \frac{\text{extract volume}}{1000} \times \text{Ca as meq/litre}$$

7.10.2 Calcium Plus Magnesium by Versenate (EDTA) Method

Magnesium in solution can be titrated with 0.01N EDTA using Eriochrome black T dye as indicator at pH 10 in the presence of ammonium chloride and ammonium hydroxide buffer. At the end point, colour changes from wine red to blue or green. When calcium is also present in the solution this titration will estimate both calcium and magnesium. Beyond pH 10 magnesium is not bound strongly to Erichrome black T indicator to give a distinct end point.

Apparatus

- Shaker
- Porcelain dish

- Beaker
- Volumetric/conical flask

Reagents

- EDTA or Versenate solution (0.01N): Same as in calcium determination.
- Ammonium chloride-ammonium hydroxide buffer solution: Same as in calcium determination.
- Eriochrome black T indicator: Take 100 ml of ethanol and dissolve 4.5 g of hydroxyl amine hydrochloride in it. Add 0.5 g of the indicator and prepare solution. Hydroxylamine hydrochloride removes the interference of manganese by keeping it in lower valency state (Mn^{2+}). Or mix thoroughly 0.5 g of the indicator with 50 g of ammonium chloride.

Procedure

1. Take 5 g air dried soil in 150 ml flask, add 25 ml of neutral normal ammonium acetate solution and shake on a mechanical shaker for 5 minutes and filter through Whatman No.1 filter paper.
2. Pipette out 5 ml of aliquot containing not more than 0.1 meq of Ca plus Mg. If the solution has a higher concentration, it should be diluted.
3. Add 2 to 5 crystals of carbamate and 5 ml of ammonium chloride-ammonium hydroxide buffer solution. Add 3-4 drops of Eriochrome black T indicator.
4. Titrate this solution with 0.01N versenate till the colour changes to bright blue or green and no tinge of wine red colour remains.

- Sodium cyanide solution (2%) or sodium diethyl dithiocarbamate crystals. This is used to remove the interference of copper, cobalt and nickel.

Calculation

If N_1 and V_1 are normality (concentration of $Ca^{2+}Mg^{2+}$) and volume of aliquot taken and N_2V_2 are the normality and volume of EDTA used respectively, then,

$$N_1V_1 = N_2V_2$$

$$\text{Or } N_1 = N_2V_2/V_1 = \frac{\text{Normality of EDTA} \times \text{Vol. of EDTA}}{\text{ml of aliquot taken}}$$

Here N_1 (Normality) = equivalents of Ca^{2+} plus Mg^{2+} present in one litre of aliquot.

$$\text{Hence, } Ca^{2+} \text{plus } Mg^{2+} \text{ meq/litre} = \frac{\text{Normality of EDTA} \times \text{Vol. of EDTA} \times 1000}{\text{ml of aliquot taken}}$$

Milliequivalent (meq) of Mg^{++} = meq ($Ca^{++} + Mg^{++}$) – meq of Ca^{++}

When expressed on soil weight basis.

Ca^{++} + Mg^{++} meq/100 g soil = 100/wt of soil x extract volume/1000 x Ca^{2+} + Mg meq/litre.

7.11 Micro-nutrients

For estimation of micronutrients also, it is the plant available form which is critical and not the total content. The major objective of soil test for micronutrients, like macronutrients, is to determine whether a soil can supply adequate micronutrients for optimum crop production or whether nutrient deficiencies are expected in crops grown on such soils. Most commonly studied micronutrients are Zn, Cu, Fe, Mn, B and Mo and the same have been dealt with here. Micronutrients are present in different forms in the soil. Among the most deficient ones is Zn, which is present as divalent cation $Zn^{2+.}$ Maize, citrus, legumes, cotton and rice are especially sensitive to zinc deficiency. Iron is present mostly in sparingly soluble ferric oxide form, which occurs as coatings of aggregate or as separate constituent of the clay fraction. Soil redox potential and pH affect the availability of iron. The form of iron that is predominantly taken up by plants is the $Fe^{2+.}$ Uptake of Fe is inhibited by phosphate levels due to the formation of insoluble iron phosphate.

Manganese, chemically behaves in the soils the same way as Fe. Soil Mn originates primarily from the decomposition of ferromagnesian rocks. It is taken up by the plants as Mn^{2+} ions, although it exists in many oxidation states. Manganese and phosphate are mutually antagonistic. Copper like zinc exists in soils mainly as divalent ions $Cu^{2+.}$ It is usually adsorbed by the clay minerals or associated with organic matter, although they have little or no effect on its availability to crops. High phosphate fertilization can induce Cu deficiency. Molybdenum mostly occurs as MoO_3, MoO_5 and MoO_2. These oxides are slowly transformed to soluble molybdates (MoO_4) which is the form taken up by plants. Boron deficiency occurs mostly in the light textured acid soils when they are leached heavily through irrigation or heavy rainfall. Different extractants have been developed for assessing plant available nutrient (element) content in the soils. The elements so extracted can be estimated quantitatively through chemical methods or instrumental techniques. Commonly used extractant for different elements are given in table 7.2.

Table 7.2 Important Soil Test methods for micro-nutrients

Element	Extractants
Zinc	EDTA + Ammonium Acetate, EDTA + Ammonium Carbonate,DTPA + $CaCl_2$, HCl, HNO_3 and Dithiozone + Ammonium Acetate
Copper	EDTA, EDTA + Ammonium Acetate, Ammonium Bicarobnate + DTPA, HCl and HNO_3
Iron	EDTA, DTPA, EDTA + Ammonium Acetate, HCl and HNO3
Manganese	Hydroquinone, Ammonium Phosphate, DTPA and EDTA + Ammonium Acetate
Boron	Hot water and Manitol + $CaCl_2$
Molybdenum	Ammonium Oxalate, Ammonium Acetate, Ammonium Fluoride and Water

7.11.1 Available Zinc, Copper, Iron and Manganese

Ethylene diamine-tera acitic acid (EDTA) + Ammonium Acetate is commonly used for extraction of many elements. DTPA (Di ethylene triamine penta acetic acid) is yet another common (universal) extractant and is widely used for simultaneous extraction of elements, like Zn, Cu, Fe and Mn. This extractant was advanced by Lindsay and Norvell (1978). Although, a specific extractant for an element which has higher correlation with plant availability may be preferred, the universal or the common extractant saves on the cost of chemicals and the time involved in estimation, especially in a service laboratory where a large number of samples need to be analysed with in a short period. Therefore, DTPA being one such extractant has been described in this publication. The estimation of elements in the extract is done with the help of Atomic Absorption Spectrophotometer (AAS). Critical limits for DTPA extractable micronutrient elements as proposed by Lindsay and Norvell (1978) are given in appendices - III.

Principle of Extraction

DTPA is an important and widely used chelating agent, which combines with free metal ions in the solution to form soluble complexes of elements. To avoid excessive dissolution of $CaCO_3$, which may release occluded micronutrients that are not available to crops in calcareous soils and may give erroneous results, the extractant is buffered in slightly alkaline pH. Triethanolamine (TEA) is used as buffer because it burns cleanly during atomization of extractant solution while estimating on AAS. The DTPA has a capacity to complex each of the micronutrient cations as 10 times of its atomic weight. The capacity ranges from 550 to 650 mg/kg depending upon the micronutrient cations.

Extracting Solution (DTPA)

DTPA 0.005M, 0.01M $CaCl_2.2H_2O$ and 0.1M TEA extractant: Add 1.967 g DTPA and 13.3 ml TEA in 400 ml distilled water in a 500 ml flask. Take 1.47 g $CaCl_2.2H_2O$ in a separate 1000 ml flask. Add 500 ml distilled water and shake to dissolve. Add DTPA+TEA mixture in $CaCl_2$ solution and make the volume to 1 litre. pH should be adjusted to 7.3 by using 1M HCl before making the volume.

Principle of Estimation

The extracted elements can be estimated by various methods, which include volumetric analysis, spectrometry and atomic absorption spectroscopy. Volumetric methods such as EDTA and $KMnO_4$ titrations are used for estimation of zinc and Mn, and iron, respectively. Copper can be estimated by titration with $Na_2S_2O_3$. Spectrometric methods are deployed in estimation of specific colour developed due to the presence of an element, which forms coloured compound in the presence of specific chemicals under definite set of conditions. The colour intensity has to be linear with the concentration of the element in question. The interference due to any other element has to be eliminated. Such methods are dithiozone method for estimation of zinc, orthophenonthroline method for iron, potassium periodate method for manganese, carbamate method for copper. The chemical methods are generally cumbersome and time taking. Hence the most commonly employed method is atomic absorption spectrometry. Here, the interference by other elements is almost nil or negligible because the estimation is carried out for an element at a specific emission spectra line. In fact in AAS, traces of one element can be accurately determined in the presence of a high concentration of other elements - IV.

7.12 Principle of Atomic Absorption Spectrophotometry

The procedure is based on flame absorption rather than flame emission and upon the fact that metal atoms absorb strongly at discrete characteristic wavelengths which coincides with the emission spectra lines of a particular element. The liquid sample is atomized. The hollow cathode lamp which precedes the atomiser emits the spectrum of the metal used to make the cathode. This beam traverses the flame and is focused on the entrance slit of a monochromator, which is set to read the intensity of the chosen spectral line. Light with this wavelength is absorbed by the metal in the flame and the degree of absorption being the function of the concentration of the metal in the flame, the concentration of the atoms in the dissolved material is determined. For elemental analysis, a working curve or a standard curve is prepared by measuring the signal or absorbance of a series of standards of known concentration of the element under estimation. From such curve, concentration of the element in unknown sample is estimated. Atomic Absorption Spectroscopy can be successfully applied for estimation of Zn, Cu, Fe and Mn. For specific estimation on AAS, hollow cathode lamps, specific to specific elements are used. The specifications of relevant hollow cathode lamps are given in appendices.

Running parameters which are specific to a particular model are given in the software provided with the equipment manual. Accordingly, the current supply, wave length of hollow cathode lamp, integration time and anticipated estimation ranges are fixed. Hollow cathode and Deutorium lamps are required to be properly aligned before starting the equipment. After proper alignment and adjustment, standard curves are prepared to ensure that the concentration of the elements in solutions perfectly relates to the absorbance.

7.12.1 Preparation of Standard Solutions

Readymade standard solutions 1000 μg/ml or 1 mg/ml (1000 ppm) of dependable accuracy are supplied with the AAS and are also available with the suppliers of chemical reagents. If the standard solutions are to be prepared in the laboratory, either metal element foils of 100% purity or the standard chemical salts can be used. The quantities of chemical required to make 1 litre standard solution of 100 μg/ml for different elements are given appendices - VI.

In case of Zn, Cu and Fe, 1000 μg/ml (1000 ppm) standard solution are preferably prepared by dissolving 1.0 g pure metal wire and volume made to 1 litre as per the method described under each element. It is diluted to obtain the required concentration. In case of Mn, Mn $SO_4.H_2O$ is preferred.

Preparation of Standard Curves

i. Zinc

Reagents

- Standard Zinc Solution: Weigh 1.0 g of pure zinc metal in a beaker. Add 20 ml HCl (1:1). Keep for few hours allowing the metal to dissolve completely. Transfer the solution to 1 litre volumetric flask. Make up the volume with glass-distilled water. This is 1000 μg/ml zinc solution. For preparation of standard curve, refer 1000 μg/ml solution as solution **A**. Dilute 1 ml of standard A to 100 ml to get 10 μg/ml solution to be designated as standard B.

- Glass-distilled or demineralized acidified water of pH 2.5 + 0.5: Dilute 1 ml of 10% sulphuric acid to one litre with glass-distilled or mineralized water and adjust the pH to 2.5 with a pH meter using 10% H_2SO_4 or NaOH. This solution is called acidified water.

Working Zn standard solutions: Pipette 1, 2, 4, 6, 8 and 10 ml of standard **B** solution in 50 ml numbered volumetric flask and make the volume with DTPA solution to obtain 0.2, 0.4, 0.8, 1.2, 1.6 and 2.0 μg/ml zinc. Stopper the flasks and shake them well. Fresh standards should be prepared every time when a fresh lot of acidified water is prepared.

Procedure

1. Flaming the solutions: Atomise the standards on atomic absorption spectrophotometer at a wave-length of 213.8 nm (Zn line of the instrument).
2. Prepare a standard curve of known concentrations of zinc solution by plotting the absorbance values on Y–axis against their respective zinc concentration on X–axis.

Precautions

- Weighing must be done on an electronic balance.
- All the glass apparatus to be used should be washed first with dilute hydrochloric acid (1:4) and then with distilled water.
- The pipette should be rinsed with the same solution to be measured.
- The outer surface of the pipette should be wiped with filter paper after use.
- After using the pipette, place them on a clean dry filter paper in order to prevent contamination.

ii. Copper

Reagents

- Standard copper solution: Weigh 1 g of pure copper wire on a clean watch glass and transfer it to one litre flask. Add 30 ml of HNO_3 (1:1) and make up the mark. Stopper the flask and shake the solution well. This is 1000 μg/ml Cu solution and should be stored in a clean bottle for further use. Dilute 1 ml of 1000 μg/ml solution of copper to 100 ml to get 10 μg/ml standard copper solutions.
- Glass-distilled or demineralized acidified water of pH 2.5 + 0.5: Same as that done for Zn.
- Working Cu standard solutions: Pipette 2, 3, 4, 5, 6 and 7 ml of 10 μg/ml standard Cu solution in 50 ml numbered volumetric flasks and make the volume with DTPA solution to get 0.4, 0.6, 0.8, 1.0, 1.2 and 1.4 μg/ml copper.

Stopper the flasks and shake them well. Prepare fresh standards every fortnight.

Procedure

1. Flame the standards on an atomic absorption spectrophotometer at a wavelength of 324.8 nm (Cu line of the instrument).
2. Prepare the standard curve with the known concentration of copper on X-axis by plotting against absorbance value on Y-axis.

iii. Iron

Reagents

- Standard iron solution: Weigh accurately 1 g pure iron wire and put it in a beaker and add approximately 30 ml of 6M HCl and boil. Transfer it to one litre volumetric flask through the funnel giving several washings to the beaker and funnel with glass-distilled water. Make the volume up to the mark. Stopper the flask and shake the solution well. This is 1.000 µg/ml iron solution.
- Glass-distilled or demineralized acidified water of pH 2.5 ± 0.5: Same as that done for Zn.
- Working Fe standard solutions: Pipette 10 ml of iron stock solution in 100 ml volumetric flask and dilute to volume with DTPA solution. This is 100 µg/ml iron solution. Take 2, 4, 8, 12 and 16 ml of 100 µg/ml solution and dilute each to 100 ml to obtain 2, 3, 8, 12 and 16 µg/ml of Fe solution.

Procedure

1. Flame the standards on an atomic absorption spectrophotometer at a wavelength of 248.3 nm (Fe line of the instrument).
2. Prepare the standard curve with the known concentration of copper on X-axis by plotting against absorbance value on Y-axis.

iv. Manganese

Reagents

- Standard Mn solution: Weigh 3.0751 g of AR grade manganese sulphate ($MnSO_4H_2O$) on a clean watch glass and transfer it to one litre flask through the funnel giving several washings to watch glass and funnel with acidified water and make the volume up to the mark. This solution will be 1 000 µg/ml Mn. A secondary dilution of 5 ml to 100 ml with acidified water gives a 50 µg/ml solution.
- Glass-distilled or de-mineralized acidified water of pH 2.5 + 0.2: Same as that for Zn.
- Working Mn standard solutions: Standard curve is prepared by taking lower concentrations of Mn in the range of 0-10 µg/ml Take 1, 2, 4, 6 and 8 ml of 50 µg/ml solution and make up the volume with DTPA solution to 50 ml to obtain 1, 2, 4, 6 and 8 µg/ml working standards.

Procedure

1. Flame the standards on an atomic absorption spectrophotometer at a wavelength of 279.5 nm (Mn line of the instrument).
2. Prepare the standard curve with the known concentration of Mn on X-axis by plotting against absorbance value on Y-axis.

7.12.2 Procedure for Extraction by DTPA

Once standard curves have been prepared, proceed for extraction by DTPA.

1. Take 10 g of soil sample in 100 ml narrow-mouth polypropylene bottle.
2. Add 20 ml of DTPA extracting solution.
3. Stopper the bottle and shake for 2 hours at room temperature (25 ^{0}C).
4. Filter the content using filter paper No.1 or 42 and collect the filtrate in polypropylene bottles
5. Prepare a blank following all steps except taking a soil sample.

Note:

The extract so obtained is used for estimation of different micronutrients. For extraction of more accurate quantity of an element which has a higher degree of correlation with plant availability, there are element specific extractants. An extractant standardized/recommended for a given situation in a country may be used. The estimation procedure on AAS, however, remains unchanged.

Estimation on AAS

1. Select an element specific hollow cathode lamp and mount it on AAS.
2. Start the flame.
3. Set the instrument at zero by using blank solution.
4. Aspirate the standard solutions of different concentrations one by one and record the readings.
5. Prepare standard curve plotting the concentration of the element concerned and the corresponding absorbance in different standard samples (as described before).
6. When the operation is performed accurately, a straight line relationship is obtained between the concentration of the element and the absorbance on AAS with a correlation coefficient which may be nearly as high as 1.0.
7. Aspirate the soil extractant obtained for estimation of nutrient element in the given soil sample and observe the readings.
8. Find out the content of the element in the soil extract by observing its concentration on the standard curve against its absorbance

Calculation

Content of micronutrient in the sample (mg/kg) = C µg/ml x 2 (dilution factor)

Where,

Dilution factor = 2.0 (Soil sample taken = 10.0 g and DTPA used = 20 ml) Absorbance reading on AAS of the soil extract being estimated for a particular element = X

Concentration of micronutrient as read from the standard curve for the given absorbance (X) = C µg/ml.

7.13 Available Boron

The most commonly used method for available B is hot water extraction of soil as developed by Berger and Truog (1939). A number of modified versions of this method have been proposed but the basic procedure remains the same. Water soluble boron is the available form of boron. It is extracted from the soil by water suspension. In the extract, boron can be analysed by colorimetric methods using reagents such as Carmine, azomethine–H and most recently by inductively coupled plasma (ICP) and atomic emission spectrometry. Colorimetric method is however, preferable due to the fact that boron being a non-metal, use of AAS for its estimation pose some limitations. The extraction method described here is the simple modification (Gupta, 1967) of the one developed by Berger and Troug (1939) in which boiling soil with water is employed.

Extraction Procedure

1. Weight 25 g of soil in a quartz flask or beaker.
2. Add about 50 ml of double distilled water and about 0.5 g of activated charcoal.
3. The mixture is boiled for about 5 minutes and filtered through Whatman Filter Paper No.42.

7.13.1 Estimation by AAS

The specifications of relevant hollow cathode lamp is given below:

Lamp current (m A°)	5
Wave length (nm)	249.7
Linear range (ìg/ml)	1-4
Slit width (nm)	0.02
Integration time (sec)	2.0

Flame Acetylene Nitrous Oxide

Running parameters which are specific to a particular model are given in the software provided with the equipment manual. Accordingly, the current supply, wavelength of hollow cathode lamp, integration time and anticipated estimation ranges are fixed. Hollow cathode and Deutorium lamps are required to be properly aligned before starting the equipment. After proper alignment and adjustment, standard curves are prepared to ensure that the concentration of the element in solutions perfectly relates to the absorbance.

Reagents

- Standard Boron Solution: Dissolve 8.819 g $Na_2B_4O_710H_2O$ in warm water. Dilute to 1 litre to get 1 000 µg/ml boron stock solution. Dilute 1 ml of standard to 100ml to get 10 µg/ml boron.
- Working standards: Take 1, 2, 3, 4, 5, 6, 7 and 10 ml of 10 ìg/ml solution anddilute each to 50 ml to get 0.2, 0.4, 0.6, 0.8, 1.0, 1.2, 1.4 and 2.0 µg/ml B.

Procedure

1. Atomise the working standards on atomic absorption spectrophotometer using acetylene-nitrous oxide as fuel instead of air acetylene fuel (as used for other micronutrients) at a wavelength of 249.7 nm.
2. Prepare a standard curve of known concentration of boron by plotting the absorbance values on Y-axis against their respective boron concentration on X axis.Measure the absorbance of the soil sample extract and find out the boron content in the soil from the standard curve.

Calculation

Content of B in the soil (µg/g or mg/kg) = C x dilution factor (10)

Where :C (µg/ml) = concentration of B as read from the standard curve against the absorbance reading of the soil solution on the spectrophotometer;

Dilution factor = 10, which is calculated as follows:

- weight of the soil taken = 25 g.
- volume of extractant (water) added = 50 ml.
- first dilution = 2 times.
- volume of the filtrate taken = 5 ml.
- final volume of filtrate after colour development = 25 ml.
- second dilution = 5 times.
- total dilution = 2 x 5 = 10 times.

7.13.2 Estimation by Colorimetric Method

The extracted B in the filtered extract is determined by the azomethine-H colorimetric method.

Apparatus

- Analytical balance
- Flask or beaker
- Volumetric flask
- Funnels
- Whatman No.42 filter paper
- Spectrophotometer

Reagents

- Azomethine-H: Dissolve 0.45 g azomethine-H and 1.0 g L-ascorbic acid in about100 ml deionized or double-distilled water. If solution is not clear, it should be heated gently in a water bath or under a hot water tap at about 300 ºC till it dissolves. Every week a fresh solution should be prepared and kept in a refrigerator.
- Buffer solution: Dissolve 250 g ammonium acetate in 500 ml deionized or double-distilled water and adjust the pH to about 5.5 by slowly adding approximately 100 ml glacial acetic acid, with constant stirring.
- EDTA solution (0.025 M): Dissolve 9.3 g EDTA in deionized or double-distilled water and make the volume upto 1 litre.
- Standard stock solution: Dissolve 0.8819g $Na_2\ B_4O_710H_2O$ AR grade in a small volume of deionized water and make volume to 1 000 ml to obtain a stock solution of 100 µg B/ml.
- Working standard solution: Take 5 ml of stock solution in a 100 ml volumetric flask and dilute it to the mark. This solution contains 5 µg B/ ml.

Procedure

1. Take 5 ml of the clear filtered extract in a 25 ml volumetric flask and add 2 ml buffer solution, 2 ml EDTA solution and 2 ml azomethine-H solution.
2. Mix the contents thoroughly after the addition of each reagent.
3. Let the solution stand for 1 hour to allow colour development. Then, the volume is made to the mark.
4. Intensity of colour is measured at 420 nm.
5. The colour thus developed has been found to be stable upto 3-4 hours.
6. Preparation of standard curve: Take 0, 0.25, 0.50, 1.0, 2.0 and 4.0 ml of 5 µg B/mlsolution (working standard) to a series of 25 ml volumetric flasks. Add 2 ml each of buffer reagent, EDTA solution and azomethine-H solution. Mix the contents after each addition and allow to stand at room temperature for 30 minutes. Make the volume to 25 ml with deionized or double-distilled water and measure absorbance at 420 nm. This will give reading for standard solution having B concentration 0, 0.05, 0.10, 0.20, 0.40 and 0.80 µg B/ml.

Calculation

Content of B in the soil (ìg/g or mg/kg) = C x Dilution factor (10)

Where,

C (µg/ml) = Concentration of B as read from the standard curve against the absorbance reading of the soil solution on the spectrophotometer.

Dilution factor = 10 which is calculated as follows:

- Weight of the soil taken = 25 g

- Volume of extractant (water) added = 50 ml
- First dilution = 2 times
- Volume of the filtrate taken = 5 ml
- Final volume of filtrate after colour development = 25 ml
- Second dilution = 5 times
- Total dilution = 2 x 5 = 10 times

Note:

1. The use of azomethine-H is an improvement over that of carmine, quinalizarin and curcumin, since the procedure involving this chemical does not require the use of concentrated acid.
2. The amount of charcoal added may vary with the organic matter content of the soil and should be just sufficient to produce a colourless extract after 5 minutes of boiling on a hot plate. Excess amounts of charcoal can result in loss of extractable B from soils.

7.14 Available Molybdenum

Molybdenum (Mo) is a rare element in soils, and is present only in very small amounts in igneous and sedimentary rocks. The major inorganic source of Mo is molybdenite (MoS_2). The total Mo content in soils is perhaps the lowest of all the micronutrient elements, and is reported to range between 0.2 µg/g and 10 µg/g. In the soil solution Mo exists mainly as $HMoO_4$ ion under acidic condition, and as MoO_4^{2-} ion under neutral to alkaline conditions. Because of the anionic nature of Mo, its anions will not be attracted much by the negatively charged colloids, and therefore, tend to be leached from the soils in humid region. Molybdenum can be toxic due to greater solubility in alkaline soils of the arid and semi-arid regions, and deficient in acid soils of the humid regions. In plants a deficiency of Mo is common at levels of 0.1 µg/g soil or less. Molybdenum toxicity (molybdenosis) is common when cattle graze forage plants with 10-20 µg Mo/g. In case of Molybdenum, ammonium acetate and/or ammonium oxalate extraction is usually carried out. Estimations can be done both by AAS and colourimetric methods with preference for the latter due to the formation of oxide in the flame in case of estimation by AAS. Therefore, chemical method has also been described. Ammonium oxalate is considered as a better extractant. However, for estimation on AAS ammonium acetate is preferred as the oxalates pose a limitation on AAS unless removed by digesting with di-acid as is described in case of colorimetric estimation.

7.14.1 Estimation by AAS Method

The specifications of relevant hollow cathode lamp

The specifications of relevant hollow cathode lamp is given below:

Lamp current (m A°)	5
Wave length (nm)	313.3
Linear range (µg/ml)	1-4

Slit width (nm) 0.02

Integration time (sec) 2.0

Flame Acetylene Nitrous Oxide

Running parameters which are specific to a particular model are given in the software provided with the equipment manual. Accordingly, the current supply, wave length of hollow cathode lamp, integration time and anticipated estimation ranges are fixed. Hollow cathode and Deutorium lamps are required to be properly aligned before starting the equipment. After proper alignment and adjustment, standard curves are prepared to ensure that the concentration of the element in solutions perfectly relates to the absorbance.

Apparatus

- Centrifuge and 50 ml centrifuge tubes
- Automatic shaker
- Atomic Absorption Spectrophotometer

Reagents

- Ammonium acetate solution (NH_4OAc) 1.0 M: Dissolve 77.09 g of ammonium acetate in 1 litre of distilled water and adjust pH to 7.0.
- Glass-distilled acidified water of pH 2.5: Same as that given under Zn estimation.
- Standard molybdenum solution: Dissolve 0.15 g of MoO_3 (molybdenum trioxide) in 100 ml 0.1M NaOH. Dilute to 1 litre to get 100 µg/ml Mo stock solution. Dilute 10 ml of the standard to 100 ml to get 10 µg/ml Mo.
- Working standard solutions: Take 1, 2, 3, 4, 5, 6, 7 and 10 ml of 10 µg/ml Mo standard solution and dilute each to 50 ml. This will give 0.2, 0.4, 0.6, 0.8, 1.0,1.2, 1.4 and 2.0 µg/ml Mo, respectively.

Procedure

1. Weigh accurately 5 g soil and transfer it into a 50 ml centrifuge tube.
2. Add 33 ml of 1M ammonium acetate solution to the tube, stopper and shake in a mechanical shaker for 5 minutes.

   Centrifuge at 2000 rpm for 5 minutes or until the supernatant is clear.
4. Decant the solution into a 100 ml volumetric flask.
5. Repeat steps 2-4.
6. Make-up the volume to 100 ml with ammonium acetate.
7. Atomise the working standards on atomic absorption spectrophotometer at a wavelength of 313.5 nm. Prepare a standard curve of known concentration of molybdenum by plotting the absorbance values on Y-axis against their respective molybdenum concentration on X-axis.

8. Measure the absorbance of the soil sample extract and find out molybdenum content in the soil from the standard curve.

Calculation

Content of Mo in the sample (mg / kg) = C μg / ml x 20 (dilution factor).

Where :

C = Concentration of Mo in the sample, as read from the standard curve for the given absorbance;

Dilution factor = 20.0(soil sample taken = 5 g and volume made to 100 ml).

7.14.2 Estimation by Colorimetric Method

Apparatus

- Spectrophotometer
- Hot plate
- Refrigerator
- Water bath

Reagents

- 50% potassium iodide solution: Dissolve 50 g in 100 ml of double-distilled water (DDW).
- 50% ascorbic acid solution: Dissolve 50 g in 100 ml of DDW.
- 10% sodium hydroxide solution: Dissolve 10 g of NaOH in 100 ml of DDW.
- 10% thiourea solution: Dissolve 10 g in 100 ml of DDW and filter. Prepare fresh solution on the same day of use.
- Toluene-3, 4-dithiol solution (commonly called dithiol): Weigh 1.0 g of AR grade melted dithiol (51 ^{0}C) in a 250 ml glass beaker. Add 100 ml of the 10% NaOH solution and warm the content upto 510 ^{0}C with frequent stirring for 15 minutes. Add 1.8 ml of thioglycolic acid and store in a refrigerator.
- 10% tartaric acid: Dissolve 10 g in 100 ml of DDW.
- Iso-amyl acetate.
- Ethyl alcohol.
- Ferrous ammonium sulphate solution: Dissolve 63 g of the salt in about 500 ml of DDW and then make the volume to one litre.
- Nitric-perchloric acid mixture (4:1).
- Extracting reagent: Dissolve 24.9 g of AR grade ammonium oxalate and 12.6 g oxalic acid in water and make the volume to one litre.

  Standard stock solution (100 ìg/ml Mo): Dissolve 0.150 g of AR grade MoO_3 in100 ml of 0.1M NaOH, make slightly acidic with dilute HCl and make the volume to 1 litre.
- Working standard solution (1 μg/ml Mo): Dilute 10 ml of the stock solution to 1litre.

Procedure

1. Weigh 25 g of air-dry soil sample in a 500 ml conical flask. Add 250 ml of the extracting solution (1:10 ratio) and shake for 10 hours.
2. Filter through Whatman No.50 filter paper. Collect 200 ml of the clear filtrate in a 250 ml glass beaker and evaporate to dryness on a water bath.
3. Heat the contents in the beaker at 500 ^{0}C in a furnace for 5 hours to destroy organic matter and oxalates. Keep overnight.
4. Digest the contents with 5 ml of HNO_3-$HClO_4$ mixture (4:1), then with 10 ml of 4M H_2SO_4 and H_2O_2, each time bringing to dryness.
5. Add 10 ml of 0.1M HCl and filter. Wash the filter paper, first with 10 ml of 0.1M HCl and then with 10 ml of DDW until the volume of the filtrate is 100 ml.
6. Run a blank side by side (without soil).
7. Take 50 ml of the filtrate in 250 ml separatory funnels and add 0.25 ml of ferrous ammonium sulphate solution and 20 ml of DDW and shake vigorously.
8. Add excess of potassium iodide (KI) solution and clear the liberated iodine by adding ascorbic acid drop by drop while shaking vigorously.
9. Add one ml of tartaric acid and 2 ml of thiourea solution and shake vigorously.
10. Add 5 drops of dithiol solution and allow the mixture to stand for 30 minutes.
11. Add 10 ml of iso-amyl acetate and separate out the contents (green colour) in colorimeter tubes/cuvettes.
12. Read the colour intensity at 680 nm (red filter).
13. Preparation of standard curve: Measure 0, 2, 5, 10, 15 and 20 ml of the working standard Mo solution containing 1 mg/litre Mo in a series of 250 ml separatory funnels. Proceed for colour development as described above for sample aliquots. Read the colour intensity and prepare the standard curve by plotting Mo concentration against readings.

Calculation

Available Mo in soil (mg/g) = A x 250/100 x 1/25 = A/ 20

Where,

A = Mo concentration in µg / ml as obtained on X-axis against a sample reading.

CHAPTER-8

Soil Biological Properties

8.1 Counting and Isolation of Soil Microbial Population by Serial Dilution Plate

Technique

Soil microbes are not visible to naked eye because of their very small size, and therefore cannot be counted as such. However, viable cells of these organisms when grown on a suitable nutrient medium multiply and give rise to cluster of cells called colonies, which are visible and can be counted. Each of these cells called colonies; develop from a single cell at appropriate dilution. Hence by enumerating these colonies which develop on a specific growth medium, the number of viable cells of a particular group of organisms in soil can be ascertained. As the number of microorganisms present in soil is very high, the soil suspension is to be sufficiently diluted to reduce the number of colonies to prevent overlapping of growth to facilitate counting. The number of colonies formed from a diluted sample multiplied by the extent of dilution (dilution factor) gives the number of viable cells present in the sample examined.

Presence of easily degradable carbonaceous substrate in rhizosphere released by metabolically active plant roots boosts the population of heterotrophic microorganisms in this zone. This phenomenon is known as "rhizosphere effect" and is expressed by the term R:S ratio; which numerically denotes the population ratios of the microorganisms in rhizosphere and bulk (non-rhizosphere) soil.

Materials

1. Soil sample (Rhizosphere and Non-Rhizosphere)
2. Sterile 1 ml/10 ml pipettes
3. Sterile water blanks 95 ml (in conical flask)

4. Sterile water blanks 9 ml (in cotton plugged test tubes)
5. Sterile petri plates

Reagents

1. Thornton's medium (for bacteria)

Potassium di-hydrogen phosphate	1.0 g
Magnesium sulphate	0.2 g
Calcium chloride	0.1 g
Sodium chloride	0.1 g
Potassium nitrate	0.5 g
Aspargin	0.5 g
Mannitol	1.0 g
Ferric chloride	Traces
Agar-agar	20g
Volume to make	1 L

2. Martin's Rose Bengal medium (for Fungi)

Dextrose	10.0 g
Peptone	5.0 g
Potassium di-hydrogen phosphate	1.0 g
Magnesium sulphate heptahydrate	0.5 g
Rose Bengal (1% solution)	3.3 ml
Agar-agar	20.0 g
Streptomycin	Traces
Volume to make	1 L

To be added after autoclaving and just before planting. Final concentration of streptomycin should be about 30 ppm

3. Ken-Knight and Munaier's medium (for Actinomycetes)

Dextrose	1.0 g
Potassium di-hydrogen phosphate	0.1 g
Sodium nitrate	0.1 g
Potassium chloride	0.1 g
Magnesium sulphate heptahydrate	0.1 g
Agar-agar	15.0 g
Volume to make	1 L

Equipment/Instrument

1. Laminar flow assembly/inoculation camber
2. Incubator
3. Autoclaves

Procedure

1. Uproot a target plant from field along with soil around its roots preferably on the day of analysis
2. Separate soil particles closely adhering to the roots by gentle tapping/ shaking. This constitute rhizosphere soil sample. The other soil sample from non-cropped area is considered non-rhizosphere sample. Avoid long period of storage and do not freeze the soil sample. If stored, keep below ambient temperature on plastic pouch to prevent drying, and incubate at room temperature for 24 hours, before analysis.
3. Prepare serial dilution from plating of the soil within the Laminar flow assembly. Add 10 g soil sample in 95 ml sterile water in 500 ml conical flask with rubber stopper. Vigorously shake to disperse soil thoroughly by giving vertical strokes (at least 20 minutes). This establishes 1:10 or 10^{-1} dilution. The shaking should be uniform from all soil samples.
4. Transfer 1 ml soil suspension obtained in previous step into 9 ml sterile water blank in the test tubes. Shake uniformly by rolling the tube between the palm of your hands to provide horizontal shaking. Continue the series in similar manner to get up to 10^{-6} dilution level. Mark the dilutions properly on the test tubes.
5. Transfer 1 ml of required dilution (10^{-4} for fungi and actinomycetes and 10^{-6} for bacteria) into sterile petri plates.
6. Pour the required medium (at approx. 45 ^{0}C) uniformly for specific organism into petri plates. Rotate clockwise and anti-clockwise to mix soil suspension with medium. Make the rotation uniform for all samples. Allow the medium to solidify. Mark details on the lid.
7. Incubate plates in inverted position at 28 ^{0}C in incubator for 2 days for fungi; 4 days for bacteria and one week for actinomycetes. Take at least four replications for each sample.
8. Examine nature of colonies developing on plates after the required incubation period. Count the number of colonies. Discard plates showing large spreaders and mold colonies. Number of colonies multiplied by the respective dilution factor will give number of viable cell g^{-1} soil.
9. Record the per cent moisture in soil sample by gravimetric method. Express results on the basis of oven dry weight of soil sample.
10. Calculate standard deviation and standard error of mean. Statistically compare the number of organisms in the rhizosphere and non-rhizosphere soil samples.
11. Calculate rhizosphere effect in terms of R:S ratio (R/S)

Precaution

The accuracy of this method depends upon the use of an appropriate selective medium, careful controlled temperature of incubation, duration of incubation and prevention of external contamination either from atmosphere or glassware/ growth medium. A large number of growth media other than mentioned in the exercise are in use, as suggested by various workers. A satisfactory medium should give reproducible results; suppress development of large spreaders and non-target organisms.

8.2 Isolation of Rhizobium from Root Nodules

Considerable amount of nitrogen from atmosphere is reduced (fixed) by rhizobia in association with legumes, and added to the soil. This takes place in tiny outgrowths called nodules formed on the root system as a result of the entry of a single cell of Rhizobium, which multiplies and grows in the nodule. Rhizobium can be isolated from the nodule and the type and strain can be identified. For isolation the nodule is surface sterilized, washed in a small amount of sterilized water and plated on a suitable growth medium. Rhizobium cells released from the crushed nodule will develop and form colonies on the growth medium, which can be purified for single cell isolation by strain identification.

Materials

1. Test tubes 5 ml sterile water
2. Petriplates (Sterilized)
3. Glass rod
4. Forceps (Sterilized)
5. Sterile tap water
6. Sterile gloves

Reagents

1. Mercuric chloride (0.1%)
2. Ethyl alcohol (70%)
3. Yeast Extract Mannitol (YEMA) with Congo Red

Equipment

1. Laminar Flow Assembly
2. Incubator
3. Autoclave

Procedure

1. Gently uproot a month old healthy leguminous plant and thoroughly wash the root system with tap water.

2. Select a plump, pink nodule on the tap root. Remove nodule with help of scalpel or a sharp razor blade along with a portion of the root at the point of its attachment.
3. Wash the nodule with the tap water several times to remove any debris adhering to it and transfer into a sterile beaker (50 ml).
4. Add a sufficient amount of 0.1% mercuric chloride solution so as to dip the nodule in it and shake by hand for 3min. Decant and then add 70% ethyl alcohol to dip the nodule in it for 230 sec. and decant the alcohol.
5. Wash the nodule with sterilized water thoroughly for 5-6 times to remove the sterilizing chemicals.
6. Place the surface-sterilized nodule in a 5 ml sterilized vial containing approx. 1 mL water. Crush with sterilized glass rod.
7. Plate the suspension of crushed nodule on Congo Red YEMA medium and incubate at 28 ^{0}C till small, round, slimy colourless/white colonies with red dot in the centre appears.
8. Transfer single isolated colony to fresh YEMA plates for purification and maintain pure culture on YEMA slant pending strain identification.

8.3 Isolation and Study of Root Surface Fungi Using Serial Root Washing Technique

Some soil fungi selectively colonize root surface closely adhering to the soil by penetrating the root cortex. There are some which grow on root surface superficially as an extension microflora. It is possible to overcome these mycelia loosely adhering to the root surface by washing with sterile water several times. The washed roots, when plated on a suitable selective medium, develop fungal colonies from the mycelium embedded in the root cortex. It is therefore, possible by this serial root washing techniques to isolate and count fungi which selectively grow on the roots.

Materials

1. Screw-capped glass vial (25 mL capacity)
2. Sterile water
3. Roots

Reagents

1. Martin's Rose-Bengal medium

Equipment /Instrument

1. Autoclave
2. Laminar flow assembly/ Inoculation chamber

Procedure

1. Wash roots thoroughly in running tap water to remove soil debris.
2. Cut roots in 2 cm long segments.

3. Transfer root segments into a sterile 25mL glass vial. Put approximately 10 mL sterile water, shake vigorously and drain the water.
4. Repeat step-3 for washing root segments 10-15 times with sterile water. Drain water completely.
5. Pour 25 ml (approx.) of Martin's Rose Bengal medium into sterilized Petri plates. Allow to solidify.
6. Place 5-6 washed root segment well spread out on the solidified medium in each petri plates. Gently press root segment with sterile forceps or glass rod.
7. Incubate for 48 h at room temperature.
8. Observe development of root surface fungal colonies.
9. Count the number of fungal colonies for all segments and calculate average number of colonies per segment.
10. Calculate number of root surface fungi per 100 cm length of root system.
11. Group root surface fungi according to colour of colonies and count for each group.
12. Attempt isolation in pure culture from a single colony developing on the root segment.

8.4 Enumeration of Soil Algae by Most Probable Number (MPN) Method

Principle

The most probable number method (MPN) for enumerating organisms in soil is based on the presence or absence of an organism in an aliquot at several consecutive dilutions in a replication, showing at least one living cell at highest dilution. It is possible to calculate, based on the probability theory, the most probable number of an organism in a sample by multiplying the results of positive growth (number of tubes) with the appropriate dilution factor. An essential prerequisite of this method is that the organism must be able to bring a recognizable change in the medium or have characteristic identifiable growth *e.g* green/blue-green pigmentation due to algal growth. Similarly this method can be used for counting nitrifying bacteria (formation of nitrate), and Rhizobium (production of nodule).

Equipment/ Instrument

Same as Serial Dilution Technique

Reagent

1. Bristol's medium

Sodium nitrate	1 g
Potassium monohydrogen phosphate	1 g
Magnesium sulphate heptahydrate	0.3 g
Calcium chloride hexahydrate	0.1 g

Sodium chloride	0.1 g
Ferric chloride hexahydrate	Traces
Volume to make	1 L

Special Supplies

1. Incubator with illumination facilities or a plant growth chamber.

Procedure

1. Weigh 10 g soil and suspend it in 95 ml sterile water in a 250 ml conical flask. Shake vigorously and allow the coarse particle to settle. This establishes 1:10 or 10^{-1} dilution.
2. Following serial dilution techniques, prepare dilution up to 10^{-6}.
3. Take 10 ml of Bristol's medium in each of the five test tubes for each level of dilution (total 30 test tubes), plug with cotton and sterilize.
4. Transfer 1 ml of the each dilution into 5 test tubes containing Bristol's medium.
5. Incubate tubes in a growth chamber (at 22 ^{0}C) or green house or near a window exposed to light for a period of one month.
6. Observe for development of green/blue-green pigmented growth and at the end count the number of tubes showing algal growth for each dilution. These are the positive tubes.
7. Determine moisture content of the soil gravimetrically.
8. Calculate the most probable number of algae in the soil sample as described below. Express the results on the basis of oven dry weight of the soil.

Calculation

1. Identify the lowest dilution in which all tubes are positive (*i.e.*, showing algal growth) and select as P_1 the number of positive tubes in that dilution (P_1 = 5 in this case).
2. Count P_2 and P_3 which represent the number of positive tubes in the next two higher dilutions. Obviously, P_2 and P_3 values will be less than P_1 (i.e., 5).
3. Refer the table of most probable number (MPN) for use with 10-fold dilution and 5 tubes per dilution as given in Cochran, W.G. (1950).
4. Find the row of numbers across the table to column headed by the observed value of P_3. The figure at the intersection is the most probable number of organism in the original sample represented in the inoculum added in second dilution.
5. Multiply this figure by appropriate dilution factor (i.e. second dilution among the three) to obtain value of original sample.

8.5 Isolation and Enumeration of Soil Protozoa

Soil protozoa directly feed (predate) on host bacteria. It will result in removal of growth of the bacterium. A series of replicated soil dilution are plated on petri dishes with growth of host bacterium. Removal of growth of host bacterium due to predation shows presence od at least one protozoa which is counted as positive (+). In case in which host bacterium is not affected, shows absence of protozoa (predator) and is counted as negative (-). By counting both (+) and (-) growth in each of the dilution and their replicates, the number of protozoa can be calculated from the total number of negative responses using Fisher's method of negative plates.

Materials

1. Soil Sample
2. Sterile 1 ml pipettes
3. Sterile test tubes
4. Glass rings (2 cm internal diameter and 1 cm depth; 1-2 mm thickness)
5. *Aerobacter aerogen*
6. Sterilized petri plates

Reagent

1. Sterile normal saline solution. Dissolve 58.5 g sodium chloride in 1000 ml distilled water. Sterilize for 15 minutes at 15 kg cm^2 pressure and 120 ^0C temperature.

Equipment/Instruments

1. Incubator

Procedure

1. Take 10 g of soil and shake for 4 min with 50 ml sterile normal saline solution.
2. From the soil suspension make a series of two fold dilution by adding 1 ml of each dilution to 1 ml sterile normal solution in each test tube. Make a series of 15 dilutions.
3. Pour about 15 ml of molten agar containing 5 g L^{-1} NaCl into each of the sterilized petri plates of 10 cm diameter.
4. Place eight sterile glass rings in each petri plate quickly before the agar solidifies.
7. Make thick suspension of *Aerobacter aerogen* from a week-old growth slant. Place one drop of this suspension to each dilution in the eight rings.
8. Place 0.5 ml suspension of each dilution in the eight rings in one petri plate. You will have thus 15 such plates with 8 rings in each.
9. Incubate plates for 15 days at 28± 1 ^0C in an incubator.
10. Examine each ring for bacterial growth. Count rings showing bacterial growth as (+) and those without bacterial growth (-) for the presence of protozoa.

11. Count total number of (+) and (-) rings. Calculate (showing no protozoa) by applying Fisher's method of negative plates from the fisher's table.

8.6 Assessing Nematode Population in Soil by Cobb's Decanting and Sieving Method.

Nematodes are microscopic soil fauna with thread like structure which are of interest as they cause some plant diseases and are often needed to be examined for their population. Besides, they also help in enhancing soil quality through i) release of nutrients from microbial biomass for plant use, ii) increase in decomposition rates and soil aggregation by stimulating bacterial activity, iii) prevention of some pathogen from establishing on plants and iv) providing prey for larger soil microorganisms. As nematodes are filamentous, they can be separated from finer soil components through a series of sieving decanting techniques. Their aggregation nature is utilized in separating them from the soil debris of equal size fraction.

Materials

1. Two plastic pans
2. A range of sieves-20, 60, 100, 200 and 325 mesh size.
3. 100 ml petri dishes.
4. Tissue paper
5. Aluminium range
6. 250 ml and 400 ml beakers.
7. Graduating counting dish
8. Baermann funnel

Equipment/Instrument

Microscope

Procedure

1. Collect soil samples from periphery of the patch where the plants show gradual decline in growth. For perennial crops, collect samples from feeder root zone. Foe annual crops samples from rhizosphere zone give better assessment.
2. Keep samples moist and store under shade. If necessary, samples may be stored at 10-15 ^{0}C without allowing the soil to dry.
3. Keep 100 g soil in pan I and soak in one litre of water for 5 minutes.
4. Break soil crumbs by hand. Discard big stones or granules separated by hand, and mix suspension thoroughly. Tilt pan and wait for 10 seconds to settle heavier particles. Then pour the muddy mixture through 20 mesh sieve into Pan II.
5. To obtain optimum yield of nematodes, add more water to the residue in Pan I, and repeat the process thrice. To ensure that even very long nematode

pass through the sieve, rinse materials on sieve with gentle jet of water or partly immerse in Pan II with gentle shaking.

6. Discard residue in Pan I and on 20 mesh sieve. Mix contents in Pan II thoroughly and pour the same way through 60 mesh sieve back to Pan I. Repeat the process two or three times and rinse the sieve as before.
7. Wash the residue on the sieve into a 250 ml beaker to observe nematode cysts.
8. Repeat the process of decanting and sieving with the rest three sieves 100, 200 and 325 mesh with the additional precaution to keep the sieve at 45° angle with ground while sieving. This will prevent the filamentous nematode to pass easily through the sieve.
9. Collect the residue on each sieve separately through backwash. If the suspension of the residue is cloudy due to presence of colloidal clay, return to the respective sieve and rinse again before proceeding to the next stage.
10. Discard materials coming through the 325 mesh. This is supposed not to carry any nematode.
11. Allow the contents of beaker to settle for 2 h and pour out excess supernatant water. Pool contents of all the beakers (except those collected over 60 mesh sieve) in 100 ml beaker and again allow to settle for reduction of the volume.
12. Pour suspension on the tissue paper in Baermann funnel assemble and leave for 24 to 48 h to allow the nematode to swim across the tissue paper in the water below leaving behind the other debris. This helps in separating the nematodes from suspension
13. Measure the volume of contents in the petri dish or collecting vial. Spread 5 to 10 per cent of it over a graduated dish for counting by observation under microscope at 40-60 x magnification.

Calculation

Express nematode population in number per unit volume of moist soil. Moreover, by measuring the initial moisture content and weight of the soil taken for measurement, the result can also be expressed on the basis of per unit weight of oven dry soil.

CHAPTER-9

Fertilizer and Manure Analysis

9.1 Detection of Adulteration in Fertilizers

Fertilizer has played a pivotal role in increasing agricultural production in any country, more so in developing nations, where the population growth rate has outstripped all other growth rates. The common fertilizers consumed in the country are Urea, Diammonium Phosphate (DAP), Single Super Phosphate (SSP), Muriate of Potash (MOP) and NPK complexes in addition of Zn Sulphate. The increasing fertilizer demand, its cost and may be, at times its scarce availability throws up many problems pertaining to making available the right quality of the fertilizers to the farmers. Adulterated fertilizers can cause serious damage to the soil and crops.

Normally two types of adulterations are found in fertilizers:

1. Misbranding of spurious/ Non-fertilizers material as high analysis fertilizers of adulteration therein.
2. Misbranding or adulteration of low grade fertilizers as high analysis fertilizers.

Common adulterants in major fertilizers are:

Name of the fertilizer	Common Adulterants
Urea	Common Salt, MOP
DAP	Clay, SSP, Rock Phosphate
SSP	Clay, Gypsum
CAN	Gypsum
MOP	Sand, Common Salt
NPK Complexes	SSP, Rock Phosphate or NPK Mixtures
Zinc Sulphate	Magnesium Sulphate

9.2 Specifications of Fertilizers

(a) STRAIGHT NITROGENOUS FERTILIZERS

1. Ammonium Sulphate

i)	Moisture percent by weight, maximum	1.0
ii)	Ammoniacal Nitrogen, percent by weight, minimum	20.6
iii)	Free Acidity (as H_2SO_4), percent by weight, maximum (0.04 for material obtained from by-product ammonia and bi-product gypsum)	0.025
iv)	Arsenic as (As_2O_3), percent by weight, maximum	0.01
v)	Sulphur (as S), percent by weight, minimum	23.0

2. Urea (46%N) (While free flowing)

i)	Moisture percent by weight, maximum	1.0
ii)	Total Nitrogen, percent by weight (on dry basis) minimum	46.0
iii)	Biuret, percent by weight, maximum	1.5
iv)	Particle size- Not less than 90 per cent of the material shall pass through 2.8 mm IS sieve and not less than 80 per cent by weight shall be retained on 1mm IS sieve	

3. Urea (Coated) 45%N) (While free flowing)

i)	Moisture percent by weight, maximum	0.5
ii)	Total Nitrogen, percent by weight, content with coating, minimum	45.0
iii)	Biuret, percent by weight, maximum	1.5
iv)	Particle size- Not less than 90 per cent of the material shall pass through 2.8 mm IS sieve and not less than 80 per cent by weight shall be retained on 1mm IS sieve	

4. Ammonium Chloride

i)	Moisture percent by weight, maximum	2.0
ii)	Ammoniacal Nitrogen, percent by weight, minimum	25.0
iii)	Chloride other than ammonium chloride (as NaCl), percent by weight,- (on dry basis), maximum	2.0
iv)	Omitted vide S.O. 1079 (E) dt.11.12.1987	

5. Calcium Ammonium Nitrate (25% N)

i)	Moisture percent by weight, maximum	1.0
ii)	Total Ammoniacal and Nitrate Nitrogen percent by weight, minimum	25.0

iii)	Ammoniacal nitrogen per cent by weight, minimum	12.5
iv)	Calcium Nitrate, percent by weight, maximum	0.5
v)	Particle size- Not less than 80 per cent of the material shall pass through 4 mm IS sieve and be retained on1mm IS sieve. Not more than 10 per cent shall be below 1 mm IS sieve	

6. Calcium Ammonium Nitrate (26% N)

i)	Moisture percent by weight, maximum	1.0
ii)	Total Ammoniacal and Nitrate Nitrogen percent by weight, minimum	26.0
iii)	Ammoniacal nitrogen per cent by weight, minimum	13.0
iv)	Calcium Nitrate, percent by weight, maximum	0.5
v)	Particle size- Not less than 90 per cent of the material shall pass through 4 mm IS sieve and be retained on1mm IS sieve. Not more than 5 per cent shall be below 1 mm IS sieve	

7. Anhydrous Ammonia

i)	Ammonia percent by weight, minimum	99.0
ii)	Water per cent by weight, maximum	1.0
iii)	Oil content by weight, maximum	20 ppm

8. Urea Super Granulated

i)	Moisture percent by weight, maximum	1.0
ii)	Total Nitrogen percent by weight (on dry basis), minimum	46.0
iii)	Biuret per cent by weight, maximum	1.5
iv)	Particle size- Not less than 90 per cent of the material shall pass through 13.2 mm IS sieve and not less than 80 per cent by weight shall be retained on 9.5 mm IS sieve.	

9. Urea (Granular)

i)	Moisture percent by weight, maximum	1.0
ii)	Total Nitrogen percent by weight(on dry basis), minimum	46.0
iii)	Biuret per cent by weight, maximum	1.5
iv)	Particle size- Not less than 90 per cent of the material shall pass through 4 mm IS sieve and be retained on 2 mm IS sieve. Not more than 5 per cent shall be below 2 mm IS sieve	

10. Urea Ammonium Nitrate (32% N) (Liquid)

i)	Total Nitrogen percent by weight, minimum	32.0
ii)	Urea Nitrogen percent by weight, maximum	16.6
iii)	Ammoniacal Nitrogen per cent by weight, minimum	7.7

iv) Nitrate Nitrogen per cent by weight, minimum 7.7

v) Specific gravity (as 15° C) 1.32

vi) Free Ammonia (as NH_3) per cent by weight, maximum 0.10

11. Neem Coated Urea

i) Moisture percent by weight, maximum 1.0

ii) Total Nitrogen percent by weight, minimum 46.0

iii) Biuret per cent by weight, maximum 1.5

iv) Benzene soluble content, per cent by weight, minimum 0.035

v) Particle size- Not less than 90 per cent of the material shall pass through 2.8 mm IS sieve and not less than 80 per cent by weight shall be retained on 1 mm IS sieve.

(b) STRAIGHT PHOSPHATIC FERTILIZERS

1. Single Superphosphate (16% P_2O_5 Powdered)

i) Moisture percent by weight, maximum 12.0

ii) Free Phosphoric Acid (as P_2O_5) percent by weight, maximum 4.0

iii) Water soluble phosphates (as P_2O_5) per cent by weight, minimum 14.5

iv) Sulphur (as S) per cent by weight, minimum 11.0

v) Neutral ammonium citrate soluble phosphate (as P_2O_5), per cent by weight, minimum 16.0

2. Rock Phosphate

i) Particle size- Minimum 90 per cent of the material shall pass w through 0.15 mm IS sieve and the balance 10 per cent of the material shall pass through 0.25 mm IS sieve. 8.0

ii) Total phosphates (as P_2O_5) per cent by weight, minimum 18.0

3. Single Superphosphate (14% P_2O_5 Powdered)

i) Moisture percent by weight, maximum 12.0

ii) Free Phosphoric Acid (as P_2O_5) percent by weight, maximum 4.0

iii) Water soluble phosphates (as P_2O_5) per cent by weight, minimum 14.0

iv) Sulphur (as S) per cent by weight, minimum 11.0

4. Superphosphate

i) Moisture percent by weight, maximum 12.0

ii) Free Phosphoric Acid (as P_2O_5) percent by weight, maximum 3.0

iii) Total phosphates (as P_2O_5) per cent by weight, minimum 46.0

iv) Water soluble phosphate (as P_2O_5), per cent by weight, minimum 42.5

5. Bone Meal, Raw

i)	Moisture percent by weight, maximum	8.0
ii)	Acid insoluble matter percent by weight, maximum	12.0
iii)	Total phosphates (as P_2O_5) per cent by weight, minimum	20.0
iv)	2 per cent citric acid soluble phosphates(as P_2O_5), per cent by weight, minimum	8.0
v)	Nitrogen content of water insoluble portion per cent by weight, minimum	3.0
vi)	Particle size- the material shall pass wholly through 2.36 mm IS sieve of which not more than 30 per cent shall be retained on 0.85 mm IS sieve.	

6. Bone Meal, Steamed

i)	Moisture percent by weight, maximum	7.0
ii)	Total phosphates (as P_2O_5) per cent by weight (on dry basis), minimum	22.0
iii)	2 per cent citric acid soluble phosphates(as P_2O_5), per cent by weight(on dry basis), minimum	16.0
iv)	Particle size- Not less than 90 per cent of the material shall pass through 1.18 mm IS sieve	

7. Rock Phosphate

i)	Particle size- Minimum 90 per cent of the material shall pass w through 0.15 mm IS sieve and the balance 10 per cent of the material shall pass through 0.25 mm IS sieve.	8.0
ii)	Total phosphates (as P_2O_5) per cent by weight, minimum	18.0

8. Single Superphosphate (16% P_2O_5 Granulated)

i)	Moisture percent by weight, maximum	5.0
ii)	Free Phosphoric Acid (as P_2O_5) percent by weight, maximum	4.0
iii)	Water soluble phosphates (as P_2O_5) per cent by weight, minimum	14.5
iv)	Particle size- Not less than 90 per cent of the material shall pass through 4 mm IS sieve and shall be retained on 1 mm IS sieve. Not more than 5 per cent shall pass through 1 mm IS sieve.	
v)	Sulphur (as S) per cent by weight, minimum	11.0
vi)	Neutral ammonium citrate soluble phosphate (as P_2O_5), per cent by weight, minimum	16.0

9. Super phosphoric Acid(70 %) P_2O_5 (Liquid)

i)	Total phosphate (as P_2O_5), per cent by weight, minimum	70.0
ii)	Poly phosphate (as P_2O_5) percent by weight, maximum	18.9

iii)	Methanol insoluble matter , per cent by weight, maximum	1.0
iv)	Magnesium (as MgO), per cent by weight, maximum	0.5
v)	Specific gravity (at 24°C)	1.96

(c) STRAIGHT POTASSIC FERTILIZERS

1. Potassium Chloride (Muriate of Potash)

i)	Moisture percent by weight, maximum	0.5
ii)	Water soluble potash content (as K_2O) per cent by weight, minimum	60.0
iii)	Sodium (as NaCl), per cent by weight(on dry basis), maximum	3.5
iv)	Particle size- Minimum 65 per cent of the material shall pass through 1.7 mm IS sieve and be retained on 0.25 mm IS sieve.	

2. Potassium Sulphate

i)	Moisture percent by weight, maximum	1.5
ii)	Potash content (as K_2O) per cent by weight, minimum	50.0
iii)	Total chlorides (as Cl) per cent by weight, (on dry basis), maximum	2.5
iv)	Sodium (as NaCl), per cent by weight (on dry basis), maximum	2.0
v)	Sulphur (as S) per cent by weight, minimum	17.5

3. Potassium Schoenite

i)	Moisture percent by weight, maximum	1.5
ii)	Potash content (as K_2O) per cent by weight, minimum	23.0
iii)	Magnesium Oxide (as MgO) per cent by weight, maximum	11.0
iv)	Sodium (as NaCl), per cent by weight (on dry basis), maximum	1.5

4. Potassium Chloride (Muriate of Potash) (Granular)

i)	Moisture percent by weight, maximum	0.5
ii)	Water soluble potash (as K_2O) per cent by weight, minimum	60.0
iii)	Sodium (as NaCl), per cent by weight, maximum	3.5
iv)	Magnesium (as $MgCl_2$) per cent by weight, maximum	1.0
v)	Particle size- Not less than 90 per cent of the material shall pass through 3.35 mm IS sieve and be retained on 1 mm IS sieve. Not more than 5 per cent shall be below 1 mm IS sieve	

5. Potash Derived from Molasses

i)	Moisture percent by weight, maximum	4.79
ii)	Total Nitrogen per cent by weight, minimum	1.66
iii)	Neutral Ammonium citrate soluble phosphate (as P_2O_5) per cent by weight, minimum	0.39
iv)	Water Soluble potash (as K_2O) per cent by weight, minimum	14.70

(d) N.P. COMPLEX FERTILIZERS

1. Diammonium Phosphate (18-46-0)

i) Moisture percent by weight, maximum 2.5

ii) Total Nitrogen percent by weight, minimum 18.0

iii) Ammoniacal nitrogen form, per cent by weight, minimum 15.5

iv) Total nitrogen in the form of urea, per cent by weight, maximum 2.5

v) Neutral Ammonium citrate soluble phosphate (as P_2O_5) per cent by weight, minimum 46.0

vi) Water Soluble phosphates (as P_2O_5) per cent by weight, minimum 41.0

vii) Particle size- Not less than 90 per cent of the material shall pass through 4 mm IS sieve and be retained on 1 mm IS sieve. Not more than 5 per cent shall be below 1 mm IS sieve.

2. Ammonium Phosphate Sulphate (16-20-0)

i) Moisture percent by weight, maximum 1.0

ii) Total Ammoniacal nitrogen form, per cent by weight, minimum 16.0

iii) Neutral Ammonium citrate soluble phosphate (as P_2O_5) 20.0

per cent by weight, minimum

iv) Water Soluble phosphates (as P_2O_5) per cent by weight, minimum 19.5

v) Particle size- Not less than 90 per cent of the material shall pass through 4 mm IS sieve and shall be retained on 1 mm IS sieve. Not more than 5 per cent shall be below 1 mm IS sieve.

vi) Sulphur (as S) per cent by weight, minimum 13.0

3. Ammonium Phosphate Sulphate (20-20-0)

i) Moisture percent by weight, maximum 1.0

ii) Total nitrogen per cent by weight, minimum 20.0

iii) Ammoniacal nitrogen per cent by weight, minimum 18.0

iv) Nitrogen in the form of Urea per cent by weight, maximum 2.0

v) Neutral Ammonium citrate soluble phosphate (as P_2O_5) per cent by weight, minimum 20.0

vi) Water Soluble phosphates (as P_2O_5) per cent by weight, minimum 17.0

vii) Particle size- Not less than 90 per cent of the material shall pass through 4 mm IS sieve and shall be retained on 1 mm IS sieve. Not more than 5 per cent shall be below 1 mm IS sieve.

viii)Sulphur (as S) per cent by weight, minimum 13.0

4. Ammonium Phosphate Sulphate Nitrate (20-20-0)

i) Moisture percent by weight, maximum 1.5

ii) Total nitrogen per cent by weight, minimum 20.0

iii) Ammoniacal nitrogen per cent by weight, minimum 17.0

iv) Nitrate nitrogen per cent by weight, maximum 3.0

v) Neutral Ammonium citrate soluble phosphate (as P_2O_5) per cent by weight, minimum 20.0

vi) Water Soluble phosphates (as P_2O_5) per cent by weight, minimum 17.0

vii) Sulphur (as S) per cent by weight, minimum 13.0

viii) Particle size- Not less than 90 per cent of the material shall pass through 4 mm IS sieve and shall be retained on 1 mm IS sieve. Not more than 5 per cent shall be below 1 mm IS sieve.

5. Ammonium Phosphate Sulphate (18-9-0)

i) Moisture percent by weight, maximum 1.0

ii) Ammoniacal nitrogen per cent by weight, minimum 18.0

iii) Neutral Ammonium citrate soluble phosphate (as P_2O_5) per cent by weight, minimum 9.0

iv) Water Soluble phosphates (as P_2O_5) per cent by weight, minimum 8.5

v) Particle size- Not less than 90 per cent of the material shall pass through 4 mm IS sieve and shall be retained on 1 mm IS sieve. Not more than 5 per cent shall be below 1 mm IS sieve.

6. Nitro Phosphate (20-20-0)

i) Moisture percent by weight, maximum 1.5

ii) Total nitrogen per cent by weight, minimum 20.0

iii) Nitrogen in ammoniacal form per cent by weight, minimum 10.0

iv) Nitrogen in nitrate form per cent by weight, maximum 10.0

v) Neutral Ammonium citrate soluble phosphate (as P_2O_5) per cent by weight, minimum 20.0

vi) Water Soluble phosphates (as P_2O_5) per cent by weight, minimum 12.0

vii) Calcium Nitrate per cent by weight, maximum 1.0

viii) Particle size- Not less than 90 per cent of the material shall pass through 4 mm IS sieve and shall be retained on 1 mm IS sieve. Not more than 5 per cent shall be below 1 mm IS sieve.

7. Urea Ammonium Phosphate (28-28-0)

i) Moisture percent by weight, maximum 1.5

ii) Total nitrogen per cent by weight, minimum 28.0

iii) Ammonical nitrogen per cent by weight, minimum 9.0

iv) Neutral Ammonium citrate soluble phosphate (as P_2O_5) per cent by weight, minimum 28.0

v) Water Soluble phosphates (as P_2O_5) per cent by weight, minimum 25.2

vi) Particle size- Not less than 90 per cent of the material shall pass through 4 mm IS sieve and shall be retained on 1 mm IS sieve. Not more than 5 per cent shall be below 1 mm IS sieve.

8. Urea Ammonium Phosphate (24-24-0)

i) Moisture percent by weight, maximum 1.5

ii) Total nitrogen per cent by weight, minimum 24.0

iii) Ammonical nitrogen per cent by weight, minimum 7.5

iv) Nitrogen in the form of Urea per cent by weight, maximum 16.5

v) Neutral Ammonium citrate soluble phosphate (as P_2O_5) per cent by weight, minimum 24.0

vi) Water Soluble phosphates (as P_2O_5) per cent by weight, minimum 20.4

vii) Particle size- Not less than 90 per cent of the material shall pass through 4 mm IS sieve and be retained on 1 mm IS sieve. Not more than 5 per cent shall be below 1 mm IS sieve.

9. Urea Ammonium Phosphate (20-20-0)

i) Moisture percent by weight, maximum 1.5

ii) Total nitrogen per cent by weight, minimum 20.0

iii) Ammonical nitrogen per cent by weight, minimum 6.4

iv) Neutral Ammonium citrate soluble phosphate (as P_2O_5) per cent by weight, minimum 20.0

v) Water Soluble phosphates (as P_2O_5) per cent by weight, minimum 17.0

vi) Particle size- Not less than 90 per cent of the material shall pass through 4 mm IS sieve and be retained on 1 mm IS sieve. Not more than 5 per cent shall be below 1 mm IS sieve.

10. Mono Ammonium Phosphate (11-52-0)

i) Moisture percent by weight, maximum 1.0

ii) Total nitrogen all in ammonical form, per cent by weight, minimum 11.0

iii) Neutral Ammonium citrate soluble phosphate (as P_2O_5) per cent by weight, minimum 52.0

iv) Water Soluble phosphates (as P_2O_5) per cent by weight, minimum 44.2

v) Particle size- Not less than 90 per cent of the material shall pass through 4 mm IS sieve and shall be retained on 1 mm IS sieve. Not more than 5 per cent shall be below 1 mm IS sieve.

11. Ammonium Nitrate Phosphate (23-23-0)

i) Moisture percent by weight, maximum 1.5

ii) Total nitrogen per cent by weight, minimum 23.0

iii) Nitrogen in ammonical form per cent by weight, minimum 13.0

iv) Nitrogen in nitrate form per cent by weight, maximum 10.0

v) Neutral Ammonium citrate soluble phosphate (as P_2O_5) per cent by weight, minimum 23.0

vi) Water Soluble phosphates (as P_2O_5) per cent by weight, minimum 20.5

vii) Particle size- Not less than 90 per cent of the material shall pass through 4 mm IS sieve and be retained on 1 mm IS sieve. Not more than 5 per cent shall be below 1 mm IS sieve.

12. Ammonium Poly- Phosphate (10-34-0) (Liquid)

i) Total nitrogen (all as ammonical nitrogen) per cent by weight, minimum 10.0

ii) Total phosphate (as P_2O_5) per cent by weight, minimum 34.0

iii) Poly phosphate (as $P2O_5$) per cent by weight, minimum 22.1

iv) Magnesium (as MgO) per cent by weight, maximum 0.5

v) Specific gravity (at 27°C) 1.4

vi) pH. 5.8-6.2

13. Ammonium Phosphate (14-28-0)

i) Moisture per cent by weight, maximum 1.5

ii) Total nitrogen per cent by weight, minimum 14.0

iii) Urea nitrogen per cent by weight, maximum 6.0

iv) Ammonical nitrogen per cent by weight, minimum 8.0

v) Neutral Ammonium citrate soluble phosphate (as P_2O_5) per cent by weight, minimum 28.0

vi) Water Soluble phosphates (as P_2O_5) per cent by weight, minimum 23.0

vii) Particle size- Not less than 90 per cent of the material shall pass through 4 mm IS sieve and be retained on 1 mm IS sieve. Not more than 5 per cent shall be below 1 mm IS sieve.

14. Diammonium Phosphate (16-44-0)

i) Moisture per cent by weight, maximum 3.0

ii) Total nitrogen per cent by weight, minimum 16.0

iii) Ammonical nitrogen per cent by weight, minimum 14.0

iv) Total Nitrogen in the form of urea per cent by weight, maximum 2.0

v) Neutral Ammonium citrate soluble phosphate (as P_2O_5) per cent by weight, minimum 44.0

vi) Water Soluble phosphates (as P_2O_5) per cent by weight, minimum 37.0

vii) Particle size- Not less than 90 per cent of the material shall pass through 4 mm IS sieve and be retained on 1 mm IS sieve. Not more than 5 per cent shall be below 1 mm IS sieve.

(e) N.P.K. COMPLEX FERTILIZERS

1. Nitro phosphate with Potash (15-15-15)

i) Moisture per cent by weight, maximum 1.5

ii) Total nitrogen per cent by weight, minimum 15.0

iii) Ammonical nitrogen per cent by weight, minimum 7.5

iv) Nitrate Nitrogen per cent by weight, maximum 7.5

v) Neutral Ammonium citrate soluble phosphate (as P_2O_5) 15.0 per cent by weight, minimum

vi) Water Soluble phosphates (as P_2O_5) per cent by weight, minimum 4.0

vii) Water Soluble potash (as K_2O), per cent by weight, minimum 15.0

viii)Particle size- Not less than 90 per cent of the material shall pass through 4 mm IS sieve and be retained on 1 mm IS sieve. Not more than 5 per cent shall be below 1 mm IS sieve.

ix) Calcium nitrate per cent by weight, maximum 1.0

2. N.P.K. (22-22-11)

i) Moisture per cent by weight, maximum 1.5

ii) Total nitrogen per cent by weight, minimum 22.0

iii) Ammonical nitrogen per cent by weight, minimum 7.0

iv) Urea Nitrogen per cent by weight, maximum 15.0

v) Neutral Ammonium citrate soluble phosphate (as P_2O_5) per cent by weight, minimum 22.0

vi) Water Soluble phosphates (as P_2O_5) per cent by weight, minimum 18.7

vii) Water Soluble potash (as K_2O), per cent by weight, minimum 11.0

viii)Particle size- Not less than 90 per cent of the material shall pass through 4 mm IS sieve and be retained on 1 mm IS sieve. Not more than 5 per cent shall be below 1 mm IS sieve.

3. N.P.K. (17-17-17)

i) Moisture per cent by weight, maximum 1.5

ii) Total nitrogen per cent by weight, minimum 17.0

iii) Ammonical nitrogen per cent by weight, minimum 5.0

iv) Urea Nitrogen per cent by weight, maximum 12.0

v) Neutral Ammonium citrate soluble phosphate (as P_2O_5) per cent by weight, minimum 17.0

vi) Water Soluble phosphates (as P_2O_5) per cent by weight, minimum 14.5

vii) Water Soluble potash (as K_2O), per cent by weight, minimum 17.0

viii) Particle size- Not less than 90 per cent of the material shall pass through 4 mm IS sieve and be retained on 1 mm IS sieve. Not more than 5 per cent shall be below 1 mm IS sieve.

4. N.P.K. (15-15-15-9 (S))

i) Moisture per cent by weight, maximum 1.5

ii) Total nitrogen per cent by weight, minimum 15.0

iii) Ammonical nitrogen per cent by weight, minimum 12.0

iv) Nitrogen in the form of urea per cent by weight, maximum 3.0

v) Neutral Ammonium citrate soluble phosphate (as P_2O_5) per cent by weight, minimum 15.0

vi) Water Soluble phosphates (as P_2O_5) per cent by weight, minimum 12.0

vii) Water Soluble potash (as K_2O), per cent by weight, minimum 15.0

viii) Sulphur (as S), per cent by weight, minimum 9.0

ix) Particle size- Not less than 90 per cent of the material shall pass through 4 mm IS sieve and shall be retained on 1 mm IS sieve.

5. N.P.K. (16-16-16)

i) Moisture per cent by weight, maximum 1.5

ii) Total nitrogen per cent by weight, minimum 16.0

iii) Ammonical nitrogen per cent by weight, minimum 8.0

iv) Nitrate Nitrogen per cent by weight, maximum 8.0

v) Neutral Ammonium citrate soluble phosphate (as P_2O_5) per cent by weight, minimum 16.0

vi) Water Soluble phosphates (as P_2O_5) per cent by weight, minimum 12.0

vii) Water Soluble potash (as K_2O), per cent by weight, minimum 16.0

viii) Particle size- Not less than 90 per cent of the material shall pass through 4 mm IS sieve and s be retained on 1 mm IS sieve.

(f) MICRONUTRIENTS

1. Zinc Sulphate Heptahydrate ($ZnSO_4.7H_2O$)

i) Matter insoluble in water, per cent by weight, maximum 1.0

ii) Zinc (as Zn) per cent by weight, minimum 21.0

iii) Lead (as Pb), per cent by weight, maximum	0.003
iv) Copper (as Cu), per cent by weigh, maximum	0.1
v) Magnesium (as Mg), per cent by weight, maximum	0.5
vi) pH not less than	4.0
vii) Sulphur (as S), per cent by weight, minimum	10.0
viii)Cadmium (as Cd), per cent by weight, maximum	0.0025
ix) Arsenic (as As) per cent by weight, maximum	0.01

2. Manganese Sulphate

i) Free flowing form	
ii) Matter insoluble in water, per cent by weight, maximum	1.2
iii) Manganese (as Mn) content per cent by weight,minimum	30.5
iv) Lead (as Pb), per cent by weight, maximum	0.003
v) Copper (as Cu), per cent by weigh, maximum	0.1
vi) Magnesium (as Mg), per cent by weight, maximum	2.0
vii) pH not less than	4.0
viii)Sulphur (as S), per cent by weight, minimum	17.0

3. Borax (Sodium Tetraborate) ($Na_2B_4O_7.10H_2O$)

i) Content of boron (as B) per cent by weight, minimum	10.5
ii) Matter insoluble in water, per cent by weight, maximum	1.0
iii) pH	9.0-9.5
iv) Lead (as Pb), per cent by weight, maximum	0.003

4. Copper Sulphate ($CuSO_4.5H_2O$)

i) Copper (as Cu) per cent by weight, minimum	24.0
ii) Matter insoluble in water, per cent by weight, maximum	1.0
iii) Soluble Iron and aluminium compounds (expressed as Fe) per cent by weight, maximum	0.5
iv) Lead (as Pb), per cent by weight, maximum	0.003
v) pH not less than	3.0
vi) Sulphur (as S), per cent by weight, minimum	12.0

5. Ferrous Sulphate ($FeSO_4.7H_2O$)

i) Ferrous iron (as Fe) per cent by weight, minimum	19.0
ii) Free acid (as H_2SO_4), per cent by weight, maximum	1.0
iii) Ferric iron (as Fe) per cent by weight, maximum	0.5
iv) Matter insoluble in water, per cent by weight, maximum	1.0

v) Lead (as Pb), per cent by weight, maximum	0.003
vi) pH not less than	3.5
vii) Sulphur (as S), per cent by weight, minimum	10.5

6. Chelated Zinc as Zn-EDTA

i) Appearance free flowing crystalline or powder or tablet	
ii) Zinc content (expressed as Zn), per cent by weight, minimum in the form of Zn-EDTA	12.0
iii) Lead (as Pb), per cent by weight, maximum	0.003
iv) pH	6.0-6.5

7. Chelated Iron as Fe-EDTA

i) Appearance free flowing crystalline/powder	
ii) Iron content (Expressed as Fe), per cent by weight, minimum in the form of Fe-EDTA	12.0
iii) Lead (as Pb), per cent by weight, maximum	0.003
iv) pH	5.5-6.5

8. Zinc Sulphate Mono-hydrate ($ZnSO_4.H_2O$)

i) Free flowing powder form	
ii) Matter insoluble in water, per cent by weight, maximum	1.0
iii) Zinc (as Zn) per cent by weight, minimum	33.0
iv) Lead (as Pb), per cent by weight, maximum	0.003
v) Copper (as Cu), per cent by weigh, maximum	0.1
vi) Magnesium (as Mg), per cent by weight, maximum	0.5
vii) Iron (as Fe) per cent by weight, maximum	1.0
viii)pH not less than	4.0
ix) Sulphur (as S), per cent by weight, minimum	15.0
x) Cadmium (as Cd) per cent by weight, maximum	0.0025
xi) Arsenic (as As) per cent by weight, maximum	0.01

9. Magnesium Sulphate

i) Free flowing –Crystalline form	
ii) Matter insoluble in water, per cent by weight, maximum	1.0
iii) Magnesium (as Mg), per cent by weight, minimum	9.6
iv) Lead (as Pb), per cent by weight, maximum	0.003
v) pH (5% solution)	5.0-8.0
vi) Sulphur (as S), per cent by weight, minimum	12.0

10. Boric Acid (H_3BO_3)

i) Boron (as B) per cent by weight, minimum 17.0

ii) Matter insoluble in water, per cent by weight, maximum 1.0

iii) Lead (as Pb), per cent by weight, maximum 0.003

(g) 100% WATER SOLUBLE COMPLEX FERTILIZERS

1. Potassium Nitrate (13-0-45)

i) Moisture per cent by weight, maximum 0.5

ii) Total nitrogen (all in nitrate form) per cent by weight, minimum 13.0

iii) Water Soluble potash (as K_2O), per cent by weight, minimum 45.0

iv) Sodium (as Na) (on dry basis), per cent by weight, maximum 1.0

v) Total Chlorides (as Cl) (on dry basis) per cent by weight, maximum 1.5

vi) Matter insoluble in water, per cent by weight, minimum 0.05

2. Mono Potassium Phosphate (0-52-34) (100% Water Soluble)

i) Moisture per cent by weight, maximum 0.5

ii) Water Soluble phosphates (as P_2O_5) per cent by weight, minimum 52.0

iii) Water Soluble potash (as K_2O), per cent by weight, minimum 34.0

iv) Sodium (as NaCl) per cent by weight, (on dry basis), maximum 0.025

3. Calcium Nitrate

i) Total nitrogen per cent by weight, minimum 15.5

ii) Ammoniacal nitrogen per cent by weight, maximum 1.1

iii) Nitrate Nitrogen (as N) per cent by weight, minimum 14.4

iv) Water Soluble Calcium per cent by weight, minimum 18.8

v) Matter insoluble in water, per cent by weight, maximum 1.5

9.4 Quick Detection of Adulteration

The "quick testing kit" developed by the Central Fertilizer Quality Control and Training (CFCL), Hyderabad is qualitative in nature which acts on the principle of physico-chemical properties of fertilizers and their adulterant like physical purity, solubility, heat treatment and chemical tests for presence of radicals present in adulterants but normally absent in fertilizers and the test of nutrients in the fertilizers.

Chemical/ glass wares required

A. Chemicals

1. Sodium Hydroxide
2. Cobalt Nitrate
3. Sodium Nitrate

4. Acetic Acid
5. Silver Nitrate
6. Ferric Chloride (Anhydrous)
7. Calcium oxide
8. Barium Chloride
9. Pot. Pyroantimonate
10. Conc. Sulphuric Acid
11. Conc. Hydrochloric Acid
12. Conc. Nitric Acid
13. Pot. Sodium Tartrate
14. Pot. Ferrocyanide
15. Copper Sulphate

B. Glassware

1. Test Tube
2. Funnel
3. Glass Rod
4. Beaker

C. Other accessories

1. Spoon
2. Filter Paper (Ordinary)
3. Magnifying Glass
4. Litmus Paper Red
5. Pestle and Morter
6. Distilled water
7. Wash Bottle

Preparation of reagents

1. **Concentrated NaOH (40%):** Take about 40g of NaOH into 100ml reagent bottle. Dissolve it in water and make the volume about 100ml.
2. **Dilute NaOH (1%):** Take 1ml of 40% of NaOH and dilute it about 40ml of distilled water.
3. **Biuret Reagent:** Take 3g of NaOH and 4gm of Pot. Sodium tartrate in a reagent bottle and dissolve it in distilled water. Add 1g of copper sulphate in the solution. Shake it well and make volume about 100ml.
4. **Silver Nitrate:** Take 1g of $AgNO_3$ and dissolve it in about 100ml of distilled water. Reagent bottle should be painted black or covered with carbon paper.

5. **Cabalti Nitrite Reagent:** Take 5g of cabalti nitrite and dissolve it in 50ml of distilled water. Add 25g of sodium nitrite and 2.5ml glacial acetic acid, shake it well and make volume about 100ml. for complete removal of the gas, the bottle should be kept open for overnight.
6. **Ferric Chloride Amm. Acetate Reagent:** Dissolve 7g of ferric chloride in a reagent bottle. Add 12g of ammonium acetate in it and make the volume about 100ml.
7. **Formaldehyde Methyl Red:** Take 100ml of 37-40% formaldehyde solution and add 1ml of methyl red indicator.
8. **Pot. Ferrocyanide:** Dissolve about 5g of pot. Ferrocyanide in 100ml of distilled water.

Method

1. Urea

Take 1g fertilizer in test tube and add 5ml of distilled water to dissolve the material. Add 5-6 drops of silver nitrate solution. Formation of white precipitate indicates that material is adulterated. The non-formation of any precipitate indicates that urea is pure.

2. DAP

a) Take 1g grind sample in a test tube and add 5ml distilled water and shake well. Add 1ml nitric acid and shake well. It it is dissolved and forms semi-transparent solution then DAP is pure. In case any insoluble material remains, then it is adulterated.

b) Take 1g grind sample in a test tube and add 5ml distilled water. Shake and filter with filter paper. Add 1ml silver nitrate solution in the filtration. Formation of yellow precipitate, which is dissolved on addition of 5-6 drops of nitric acid conforms presence of phosphate in the material.

c) Mix small quantity of lime in sample and rub. Smell of pungent ammonia gas indicated presence of nitrogen.

3. MOP

a) Take 1g fertilizer in test tube and 5ml distilled water and shake well. Most of the fertilizer is dissolved and a few insoluble particles float on the surface of water. This confirms the purity of MOP. However, in case more undissolved material is lying in the bottom of the test tube, this indicates adulteration in fertilizer.

b) Take 1g fertilizer and add 10ml distilled water. Shake it well and filter. Add 5-6 drops of cobalt nitrate reagent in the filtrate. Formation of yellow precipitate indicates the presence of potassium in the fertilizers. Non formation of the precipitate indicates that the material is spurious.

4. *NPK Complexes*

a) Take 1g fertilizer in test tube and 5ml distilled water. Add 1ml concentrated NaOH solution and heat the tube from the side. Place moist red litmus paper on the mouth of tube. Change of red litmus paper into blue colour confirms the presence of nitrogen. No change in the litmus paper colour indicates absence of nitrogen and fertilizer may be considered as suspected.

b) Take 1g fertilizer and add 5ml distilled water. Shake well and filter. Add 2ml of ferric chloride-ammonium acetate reagent. Formation of yellow precipitate, which gets dissolved in 5-6 drops of conc. Nitric acid, confirms the presence of phosphate. Non formation of yellow precipitate indicates absence of phosphate.

c) Take 1g fertilizer and add 5ml distilled water. Shake well and filter. Add 2ml of formaldehyde solution and wait for 5mins, the colour of solution changes to red. Add dilute NaOH drop by drop so that the colour changes to yellow. Add 1ml cobalt nitrite reagent. Formation of yellow precipitate indicates the presence of potash in the sample.

5. SSP

Take 1g fertilizer in test tube and add 5ml distilled water. Shake well and filter. Add 1-2 drops 2% ammonium hydroxide solution. Add 5-6 drops of silver nitrate solution. The formation of yellow precipitate indicates the presence of phosphate. On addition of 5-6 drops of nitric acid on the precipitate, a milky solution is formed. The non-formation of yellow precipitate indicates absence of phosphate.

6. Zinc sulphate

Take 1g of fertilizer in test tube and add 5ml of distilled water shake well and filter through filer paper. Add about 1-10 drops of NaOH solution. A white jelly like precipitate is formed. Then add 10-12 drops of conc. NaOH. If the precipitate is dissolved, it confirms the purity of material. In case the precipitate is not dissolved, this indicates that the material is adulterated with magnesium sulphate.

7. Copper sulphate

Take 1g of fertilizer in test tube and add 5ml of distilled water. It forms a transparent blue colour solution. Add 1ml of Pot. ferrocyanide solution, the formation of brown precipitate indicates the presence of copper sulphate.

8. Detection of misbranding or adulteration of SSP or Gypsum granules as DAP

Take 1g grind fertilizers in test tube. Add 5ml of distilled water. Add 1ml conc. Nitric acid. If the material is completely dissolved and gives semi transparent solution, it indicates pure DAP. However, if the insoluble material remains at bottom of test tube with muddy suspension, it indicate adulteration with SSP or gypsum in DAP.

Physical Test

The physical test include shape and size of particles, presence of foreign materials and solubility of material in water and as every fertilizer has specific characteristics, observation different from normal sample creates suspicion about quality. Some fertilizers give cold effect when dissolved in water like urea, MOP and NPK complexes. The presence of adulterants reduces the effect.

9.5 Moisture Determination in Manures and Fertilizers

Determination of moisture in manures and fertilizers is very important because its excessive presence not only results in nutrient dilution but also leads to lump and aggregate formation, particularly in fertilizers, which may adversely affect the handling and application in field. Moisture in fertilizers may be of two kinds: (a) free water (b) bound moisture as water of crystallization. Bound water is present in many fertilizers such as single superphosphate [$3Ca(H_2PO_4).H_2O+7CaSO_4$], triple superphosphate [$Ca(H_2PO_4).H_2O$], Zinc sulphate ($ZnSO_4.7H_2O$ and other hydrated fertilizers. For moisture determination, physical and chemical methods are used. Physical methods include (i) oven drying and (ii) vacuum desiccation method, while chemical methods include Karl Fischer Reagent method. Depending on the thermo stability of the fertilizers an appropriate method of moisture determination is selected.

9.5.1 Physical Methods

9.5.1.1 Oven Drying

Principle-: All manures and few fertilizers such as ammonium sulphate [$(NH_4)_2SO_4$], muriate of potash [KCl] and sulphate of potash [K_2SO_4] are thermo-stable, that is these do not release any volatile substances other than water on drying. The moisture content in them can be conveniently determined by a simple method *i.e* by oven drying, A weighed quantity of manure/fertilizer material is dried in an oven at 100 ± 2 ^{0}C for 3-4 h or till it attains constant weight. For ammonium sulphate and potassium salts drying the sample at 130 ± 2 ^{0}C are recommended. After attaining constant weight, the moisture in the sample is calculated as percentage loss in weight and reported on fresh weight basis.

Procedure

Take about 10 g of manure/fertilizer material in a moisture box or porcelain dish or watch glass of known weight and record the weight of the material. Let it be W_1 g.

Dry it an oven at 100 ± 2 ^{0}C (130 ± 2 ^{0}C in case of sodium nitrate, ammonium sulphate and potassium salts) till it attains a constant weight. This means that the consecutive weights should be same.

Record the dry weight of the material. Let it be W_2 g (9.5 g)

Calculate the weight of the moisture present as the difference of the wet and dry weights. This comes out to be (W_1-W_2) g.

Calculations

Let the weight of moist sample be W_1 (10.0) g and the weight of dry sample W_2 (9.5)g

Then, moisture = $(W_1\text{-}W_2)$ g = (10.0-9.5) g = 0.5 g

Therefore, the percent moisture (on fresh weight basis) (%) = $[(W_1\text{-}W_2)/W_1] \times 100$

= (0.5 /10) x100 = 5.0%

While, the per cent moisture on dry weight basis (%) = $[(W_1\text{-}W_2)/W_2] \times 100$

(0.5/9.5) x 100 = 5.26%

Thus it can be seen from the above example that the percentage on dry basis is slightly higher than that on wet basis. It may noted here that oven drying method is applicable to manures and only to thermally stable fertilizers such as ammonium sulphate, muriate of potash and potash of sulphate and is not applicable to ammonium chloride, urea, diammonium phosphate(DAP) and monoammonium phosphate (MAP), *etc.*

9.5.1.2 Vacuum Desiccation Method

Principle: This method is used for determination of moisture in thermo-unstable fertilizers materials. The fertilizer materials, which fall under this category are, urea, (NH_2CONH_2), ammonium chloride (NH_4Cl), ammonium nitrate (NH_4NO_3), calcium ammonium nitrate(CAN), potassium nitrate (KNO_3), single superphosphate (SSP), triple superphosphate (TSP), *etc.* In thermo-unstable fertilizer materials, moisture cannot be determined by oven drying method because at higher temperature the maerial will decompose and volatile substances other than moistuire will be released e.g Upon heating ammonium chloride sublimes; urea, DAP, MAP release NH_3. Therefore moisture in thermo-unstable fertilizers can be determined either by vacuum desiccation or by following the chemical methods (Karl Fischer Reagent).

The sample is kept in the vacuum desiccator for 24 h or till it attains constant weight. The percentage moisture can be determined by application of the formula given under the above section dealing with oven drying method.

Procedure

- Weight out 10.0 g (approx.) of the manure or fertilizer material in a moisture box or porcelain dish of known weight. Record the exact weight. Let it be W_1 g
- Dry the material in a vacuum desiccators containing previously boiled Conc. H_2SO_4 or dry fused $CaCl_2$ for 24 h or till it attains a constant weight. Let the dry weight be 9.7 (W_2) g.

Calculation

Let the weight of moist sample is W_1 g and the weight of dry sample is W_2 g

Then the amount of water in the material is = $(W_1\text{-}W_2)$ = (10.0-9.7) g = 0.3 g

Thus, percentage of moisture on fresh weight basis is = $[(W_1\text{-}W_2)/W_1] \times 100$

= (0.3/10.0) x 100 = 3.0%

But, percentage of moisture on dry weight-basis is = $[(W_1\text{-}W_2)/W_2] \times 100$

= (0.3/9.7) x100 = 3.09%

9.5.2 Chemical Method-Karl Fischer Reagent Method

Principle: Karl Fischer Reagent (KFR) method is a chemical method of determination of moisture. This method of moisture determination is used in volatile substances or thermo-unstable fertilizers or urea, urea based fertilizers and DAP. Vacuum desiccation method can also be used but this procedure is time consuming. For rapid and accurate determination of moisture in such fertilizers, chemical method is preferred. This method involves titration of water against KFR as described below.

Karl Fischer Reagents consists of four ingredients namely i) anhydrous iodine (I_2) ii) anhydrous sulphur dioxide (SO_2), iii) anhydrous pyridine (C_5H_5N), and anhydrous methanol(CH_3OH). Chemical reactions involved and role of each ingredientare as under:

Major reaction takes place between I_2 and H_2O present in the fertilizer

$$I_2 + H_2O \rightarrow 2HI + O$$

To make the equilibrium move in forward direction, the nascent oxygen generated has to be removed. The sequence of reactions, which causes the forward movement of equilibrium, is :

$$SO_2 + O \rightarrow SO_3$$

SO_2 is a reducing agent and therefore captures nascent O liberated and gets converted to SO_3. Since SO_3 can react with H_2O to form H_2SO_4, this also has to be eliminated from the sphere of reaction. This role is played by pyridine according to the following reaction. The end product of the reaction is pyridinium methyl bisulphate.

$$C_5H_5N + SO_3 \rightarrow C_5H_5NSO_2O^-$$

(Pyridinium sulphonate)

Here, N acts as electron donor. Therefore, the change appears on N

$$C_5H_5NSO_2O^- + CH_3OH \rightarrow C_5H_5N\text{-}CH_3\text{-}HSO_4$$

(Pyridinium methyl bisulphate)

Reagents

Kark Fischer Reagent

Anhydrous methanol (CH_3OH)

Karl Fischer Apparatus: The instrument consists of two burettes, one for the reagent and other for the standard solution of water in methanol. Each burette is attached to a reservoir, which may hold up to 1 litre of liquid and a series of guard tubes to prevent the entry of atmospheric moisture. The titration vessel is fitted with an air-tight cover, and is provided with a pair of platinum electrodes connected to micro-ammetre. Provision is made for stirring the contents of the vessel of a magnetic stirrer. The scale of micro-ammetre is often marked with "Excess reagent" and "Excess water" sign. The usual experimental procedure is to add a slight of the reagent (KFR) so that all the water in the sample under the test is reacted and the excess KFR is back titrated with the standard water in methanol solution.

Procedure

The procedure for the determination of moisture with KFR consists of two steps i) standardization of KFR ii) titration of moisture in sample with standard KFR

i) *Standardization of KFR with disodium tartrate* ($Na_2C_4O_6.2H_2O$)

- The original Karl Fischer Reagent (KFR) is prepared with an excess of methanol which is somewhat unstable and required frequent standardization. This may be done with pure disodium tartrate ($Na_2C_4O_6.2H_2O$), which contains 15.66% water, or, more commonly by means of a solution of water in methanol.
- Pipette out 25 ml of absolute methanol (CH_3OH) into a dry titration vessel and titrate it aginst KFR. The end point will be marked by a color change from brownish yellow to amber due to presence of unreacted iodine.
- Add accurately 0.5 (W1) g of pure disodium tartrate (15.66% water) and stir it thoroughly.
- Titrate the contents against with the KFR reagent. Note the volume of KFR consumed and mark it as V_1.

ii) *Titration of moisture in fertilizer material with standard KFR*

- Take 25 ml of methanol into a dry titration vessel and quickly transfer to it an exactly weighed quantity of urea (or any other fertilizer material).
- Shake the contents of the flask well to ensure homogeneity and titrate to the end point. Let this volume be V_2.
- Calculate the water present in the Karl Fischer Reagent from the formula given below.

Calculation

It is noted that the pure disodium tartrate dihydrate contains 15.665 water

Therefore, 100 g of disodium tartrate dihydrate contains 15.66%

Therefore W1 g of disodium tartrate dihydrate contains 15.66 x W_1/100 g of water

$$= 0.1556 \text{ g} \times W_1 \text{ g water}$$

Suppose, the volume of KFR reagent used for Standardization is V_1 mL

Therefore V_1 ml of KFR reagent = 0.1556 x W_1 g water

Therefore 1 ml of KFR reagent = 0.1556 x W_1/V_1 g water

Now, suppose the volume of KFR reagent used for the W2 g of fertilizer material is V_2 mL

Then, 1 mL of KFR reagent = 0.1556 x W_1/ V_1 g water

V_2 mL of KFR reagent = 0.1556 x W_1 XV_2/V_1 g water

Thus, W_2 g fertilizer material contains = 0.1556 x W_1x V_2 /V_2 g water

1 g fertilizer material contains = $(0.1556 \times W_1 \times V_2)/(V_1 \times W_1)$ g water

Therefore, 100 g fertilizer material contains = $(0.1556 \times W_1 \times V_2 \times 100)/(V_1 \times W_1)$ g water

Therefore, moisture per cent in sample by weight = = $(15.66 \times W_1 \times V_2)/(V_1 \times W_1)$

Note/Precautions

- Since the accuracy of the result depends upon the exact determination of end point of the reaction, which can either be determined electrometrically or visually, care should be taken to ensure it.
- The color change at the end point will be from brownish yellow to amber. If visual detection is difficult then electrometric method can be followed.
- This method can be used for determination of moisture in both solid and liquid fertilizers
- All the apparatus should be free from moisture.

9.6 Methods for Nitrogen Estimation-Determination of total Nitrogen in fertilizers and Manures

Most of nitrogen present in manure sample is in organic form. Mannures also contain variable amount of NH_4 and NO_3^--N. On the other hand, N in fertilizer may be present as i) amide (e.g urea), ii) ammonium [e.g $(NH_4)_2SO_4$, NH_4Cl], iii) nitrate [e.g. $Ca(NO_3)_2$, $NaNO_3$, KNO_3], iv) ammonium and nitrate form [NH_4NO_3, Calcium ammonium nitrate (CAN). The procedure to be followed for N determination in the fertilizer sample, thus depends on the form of N expected to be present in it.

9.6.1 Determination of total N Excluding Nitrates (Applicable for Manures, Urea, Urea-based Fertilizers, Complex NPK Fertilizers, DAP, etc)

Principle: Total N in manures and fertilizers free from NO_3^--N is determined by Kjeldahl method. Some pre-treatment is required for reducing nitrate (if present) into ammonium form before digestion. Organic form of N present in few fertilizers and all the manures has to be converted into inorganic (ammonium) state by a procedure known as Kjeldahl digestion –distillation method. This is essentially a two step process. In the first step, the sample is digested with concentrated H_2SO_4 using a digestion mixture. In the second step, the acid digest is distilled with 40% of NaOH and the NH_3 thus liberated is absorbed in either 4 % boric acid or standard acid (HCl or H_2SO_4).

The original Kjeldahl procedure has been modified many times. The various modifications differ from the original in the addition of K_2SO_4 or Na_2SO_4 to raise the temperature of the digest and the use of other catalyst of mercury (Hg or HgO). It is subjected to many difficulties, any of which leads to low results. For example, the whole of organic N is not converted to $(NH_4)_2SO_4$ until some considerable time is allowed for the digestion. Thus an hour of digestion may be necessary to turn the digest depending on the catalyst used and at the end of digestion it becomes clear solution. To release the entire N present in the sample, then most important consideration being the catalyst selected and the digest temperature employed. If the temperature is too low (below 360 ^{0}C), the release of N is slow or incomplete and if too high (over 410 ^{0}C), some of N is lost as NH_3.

Digestion mixture is used in order to raise the temperature during the digestion and to shorten the digestion time. A mixture of K_2SO_4 and $CuSO_4$ powder (a catalyst) in ratio of 10:1 is normally used. Selenium powder, another catalyst is more efficient tan $CuSO_4$ since it shortens the time of digestion required to obtain maximum values for N. Some catalyst prefers a mixture of $CuSO_4$ and selenium (20:1). Thus, a mixture of K_2SO_4, $CuSO_4$, and Se powder in the ratio of 200:20:1 can be conveniently used.

A known quantity of representative sample is digested with concentrated H_2SO_4 using a digestion mixture to bring the organic N into inorganic form of N (NH_4-N). It is then distilled as per the Kjeldahl method, which is employed in macro, micro or ultra-micro scale. In Kjeldahl method, organic N is digested with H_2SO_4 and nitrogen is converted to $(NH_4)_2SO_4$. Organic N when digested with H_2SO_4 is oxidized to CO_2 and H_2O and their inorganic N is released. During digestion, part of H_2SO_4 is reduced to SO_2, which in turn reduces N- materials to NH_3. Ammonia then combines with H_2SO_4 and converts $(NH_4)_2SO_4$ at the end of digestion.

$$\text{Organic N+}H_2SO_4 \xrightarrow{\text{(oxidized)}} CO_2(\uparrow)\text{+ } H_2O_2 \text{ +inorganic-N}$$

$$H_2SO_4 \xrightarrow{\text{(reduced)}} SO_4$$

$$\text{N-materials} + SO_2 \longrightarrow NH_3$$

$$2NH_3 + H_2SO_4 \longrightarrow (NH_4)_2SO_4$$

Determination of total N including nitrates (Applicable for manures and fertilizers containing organic N and nitrate N)

Principle: If NO_3^- -N is present, it is to be converted into NH_4^+-N by reduction. Otherwise it will react with conc. $H_2SO_4$and will be lost as HNO_3. To prevent this loss of NO_3^- it is first changed to NH_4^+ form by reduction with sodium thiosulphate ($Na_2S_2O_3$) and salicylic acid (C_6H_4-OH-COOH). The NO_3^- is first converted into HNO_3 in the presence of H_2SO_4, which is then fixed by salicylic acid in the form of nitrosalicylic acid. Nitrosalicylic acid is then converted into amionosalicylic acid by the action of reducing agent i.e. sodium thiosulphate ($Na_2S_2O_3$). The amionosalicylic thus formed is digested with H_2SO_4 and digestion mixture K_2SO_4, $CuSO_4$, and Se powder has been explained in the preceding section. Depending upon the quantity and type of material, the digestion process is likely to take about 1 to 2 hours.

The sequence of reactions taking place may be represented as:

$$2NO_3^- + H_2SO_4 \longrightarrow 2HNO_3 + SO_4^{2-}$$

$$C_6H_4\text{-OH-COOH} + HNO_3 \longrightarrow C_6H_3\text{-}NO_2\text{-OH-COOH} + H_2O$$

$$C_6H_3\text{-}NO_2\text{-OH-COOH} + 6H^+ \longrightarrow C_6H_3\text{-}NH_2\text{-OH-COOH} + 2H_2O$$

(*p*-nitrosalicylic acid) (*p*-aminosalicylic acid)
(coming from $Na_2S_2O_3$)

$$2C_6H_3\text{-}NH_2\text{-OH-COOH} + H_2SO_4 + 11O_2 \longrightarrow (NH_4)_2SO_4 + 12CO_2 + 4H_2O$$

At the end of digestion, a clear solution of acid digest is obtained where nearly all nitrogen is converted into $(NH_4)_2SO_4$ form. It is then distilled with 40% NaOH solution. During distillation NH_3 is liberated due to reaction given below

$$(NH_4)_2SO_4 + 2NaOH \longrightarrow Na_2SO_4 + 2NH_3 + 2H_2O$$

Ammonia evolved may be absorbed in excess HCl or H_2SO_4 (0.1N) containing 2-3 drops of methyl red indicator. The unreacted acid is then back titrated with standard (0.1N) NaOH to find out the total N content of the sample.

Alternatively, the liberated NH_3 can be absorbed in 4% boric acid (H_3BO_3) containing mixed indicator (pH 4.5) and titrated with standard acid (0.1 N or 0.02N) to find out the total N content of thje sample. The reactions can be represented as

$$2NH_3 + H_2SO_4 \longrightarrow (NH_4)_2SO_4$$

$$NH_3 + H_3BO_3 \longrightarrow NH_4H_2BO_3$$

$$2NH_4H_2BO_3 + H_2SO_4 \longrightarrow (NH_4)_2SO_4 + 2H_3BO_3$$

Reagents

i) Concentrated H_2SO_4

ii) Digestion mixture. Mix K_2SO_4 and $CuSO_4$ in the ratio of 1:10 in a porcelain pestle and grind with mortar to make homogenous powder.

iii) Methy red indicator: This can be prepared by dissolving 30 mg of methyl red powder in 100 ml of 95% ethyl alcohol.

iv) Mixed indicator) methyl red and bromocresol green): Mixed indicator is prepared by dissolving 0.066 g methyl red and 0.099 g brmocresol green in 100 ml of ethyl alcohol.

v) 4 % boric acid (H_3BO_3) solution (pH 4.5): Dissolve 40 g H_3BO_3 in approximately 900 ml hot distilled water in one litre volumetric flask. Add 20 ml of ml of mixed indicator to it. 4 % H_3BO_3 has pH of 4.0, which should be adjusted to 4.5 by adding dilute NaOH or HCl solution using a pH meter. Finally, make up volume to 1 L. The final colour of boric acid is bluish purple.

vi) 0.1N NaOH solution: Dissolve 4.0 g of NaOH in water and make up volume to 1 L.

Standardize it against 0.1 N potassium hydrogen phthalate (primary standard).

vii) 0.1 N H_2SO_4: Dissolve 2.78 ml of concentrated H_2SO_4 in 1 L. water and make up volume to 1L. Standardise this with sodium carbonate (0.1 N Na_2CO_3) solution or a standard alkali solution (KOH or NaOH).

viii) NaOH (40%) solution: Dissolve 400 g of NaOH flakes in 1 litre distilled water and allow it to cool before use.

ix) salicylic acid (for samples containing NO_3^-N): Dissolve 25 g of salicylic acid in 1L of conc. H_2SO_4.

Procedure:

- Place 1 g fertilizer or manure sample in 250 ml in nitrogen digestion and distillation flask (micro-Kjeldahl flask).
- Add 2-3 g of digestion mixture .Add 0.5 g of salicylic acid if the material contains nitrate-N.
- Add 8-10 ml of concentrated H_2SO_4 and digest the contents of flask in a digestion block setting the temperature at about 370-410 ^{0}C.
- Open the water tap connected with the digestion block for scrubbing the acid fumes during the digestion. Continue the digestion for 1-2 h or till a green colour solution is obtained.
- Take out the digestion tube from the digestion block and allow it to cool to room temperature.
- Place the distillation tube properly in the nitrogen distillation unit properly.
- Pipette out 20-25 ml of boric acid containing mixed indicator or 0.1 N H_2SO_4 in a 250 ml of conical flask. Place the conical flask in the distillation unit in such a way that the outlet of the condenser tube remains well in the acid.
- Add about 50-60 ml of 40% NaOH solution to micro-Kjeldahl distillation tube.
- Adjust the distillation time say, 3 minutes.
- Run the instrument and collect the distillate by absorbing liberated NH_3 either in 4% boric acid (or excess 0.1 N H_2SO_4)
- Titrate the liberated NH_3 absorbed in boric acid against standard H_2SO_4 (0.1N) solution or back titrated the un-reacted H_2SO_4 added against 0.1 N NaOH solution.

Calculation

- 1000 ml of 1 N H_2SO_4 = 14 g of N
- Or 1 ml of 1 N H_2SO_4 = 14 mg of N
- 1 ml of 0.1 N H_2SO_4 = 1.4 mg of N
- Let the volume of 0.1 N H_2SO_4 consumed in the reaction

- Then the solution contains V_1 x 1.4 mg N or 0.0014 x V_1 g N
- Let, W be the weight (g) of the material taken for digestion
- Then, the percentage of N consumed in the material = (0.0014 xV_1 x 1000)/ W = 0.14 xV_1 /W_1)
- When the boric acid is used for absorbing for NH_3, the per cent N can also be calculated from the formula given below
- Percentage of N in fertilizer or manure (%) = [(T-B) x N x 1.4]/S
- Where T = volume (ml) of standard H_2SO_4 required for sample
- B = volume (ml) of standard H_2SO_4 required for blank
- N = Normality of acid
- S = weight of sample taken

Note/Precaution

- Digestion should be carried out at temperature range of 370-410 ^{0}C for completeness of digestion. Temperature below 370 ^{0}C will result incomplete digestion and temperature above 410 ^{o}C will cause nitrogen loss in form of NH_3.
- The salt: acid ratio (weight: volume) should be less than 1:1at the end of digestion.
- The material after digestion should not solidify. To prevent it add 20-25 ml of distilled water to the acid digest.
- Neither the volume or strength of boric acid need to be known exactly because the NH_3-borate is titrated back to H_3BO_3 in the titration with standard acid (HCl or H_2SO_4)

9.6.2 Estimation of Urea-Nitrogen by Hydrolysis Method

Principle: Urea is an organic compound and it happens to be first organic compound synthesized *in vitro*. The total N in this compound can be determined either by i) by digestion with concentrated H_2SO_4 using digestion mixture containing K_2SO_4 or Na_2SO_4 and $CuSO_4$ + Se powder as per the method given for determination of total N earlier or ii) by hydrolyzing urea molecule with urease enzyme . The enzyme is extracted from protein rich seeds such as soyabean, pigean pea, jack beans or any other suitable sources. Alternatively, the urease enzyme tablets produced by pharmaceuticals industry can be used. The method of urea-N determination depends upon the property of urea to get hydrolysed in the presence of urease enzyme. The method consists of two steps i) Hydrolysis of urea with urease enzyme ii) Titration of $(NH_4)_2CO_3$ formed as result of hydrolytic breakdown of urea.

The urea solution undergoes hydrolysis and $(NH4)_2CO_3$ is formed in accordance with the following equation

$$NH_2CONH_2 + 2H_2O \xrightarrow[\text{Hydrolysis}]{\text{Urease}} (NH_4)_2\,CO_3$$

Ammonium carbonate formed can be titrated with standard acid (0.1 NHCl) to calculate the quantity of urea-N released due to hydrolysis in the solution.

$$(NH_4)_2CO_3 + 2\,HCl \longrightarrow 2NH_4Cl + CO_2 + H_2O$$

Reagents

i) Urease enzyme: Soak about 100 g seeds in sufficient amount of water and keep it overnight. When the seeds become soft macerate it with water in a porcelain pestle and mortar and filter through clean piece of cloth or filter (Whatman No 1).

ii) Urea solution (20%): Dissolve 20 g urea in 100 ml of water

iii) Standard HCl solution (0.05 N)

iv) Methyl red indicator

Procedure

Pipette out 20 ml of 20% urea solution in a 250 ml conical flask

Add 5 ml of the soyabean extract or 5 g of paste to urea solution and keep it overnight.

After 24 h titrate the content against standard HCl using 2-3 drops of methyl red indicator.

Note down the amount of acid consumed and calculate the urea-N as per the calculation given below

2 moles of HCl = 1 mole of $(NH_4)_2CO_3$ =1 mole of urea

or 2000 ml of 1 N HCl = 60 g of urea = 60 x 1000 mg urea

or,1 ml of 1 N HCl = 3 mg of urea

On the other hand, 1 mL of 1 N HCl = 14 mg of N

Therefore, X ml of Y N HCl = (14 x X x Y) mg of N

If V, is volume (ml) of 0.1 N HCl consumed for titration

Then, the quantity of urea present in the solution is = V x 3mg = 0.003 x V g

Let, W g is the weight of the material taken

Then, the percentage of urea in the material (%) = (0.0 x V/W) x 100%

9.7 Phosphorus Determination in Fertilizers

Gravimetric analysis is a technique that involves the determination of the mass of a chemical species of known composition that can be related to the analyte by mass stoichiometry. The most common gravimetric technique involves precipitation. In this experiment, the water-soluble phosphate in fertilizer is precipitated from solution with magnesium sulfate. The precipitate is magnesium ammonium phosphate hexahydrate.

The chemical reaction equation may be written as

$PO_4^{3-}(aq) + Mg^{2+}(aq) + NH_3\ (aq) + 7\ H_2O\ (l) \rightarrow MgNH_4PO_4{\cdot}6H_2O(s) + OH^-(aq)$

The sulfate ion is a spectator ion and is not included in the chemical reaction equation. Based on the mass of fertilizer dissolved and the fertilizer analysis, you will need to calculate the volume of $MgSO_4$ solution needed to form your precipitate. The solution is made basic with aqueous ammonia in order to precipitate the entire mass of phosphate ion. If the solution is not sufficiently basic, the following reaction occurs, and some of the phosphate is not precipitated, which produces inaccurate results:

$PO_4^{3-}(aq) + H_3O^+(aq) \rightarrow H_2PO_4^{2-}(aq) + H_2O\ (l)$

The precipitate must be dried at room temperature to prevent the waters of hydration from being lost. Once the mass of $MgNH_4PO_4{\cdot}6H_2O(s)$ is obtained, the stoichiometric ratios between P, P_2O_5, and the product will yield the amount of P_2O_5 in the fertilizer.

Procedure

Weigh between 3 and 5 g of fertilizer.

Record the brand and type of fertilizer and the percentage analysis. Place in a 1000 ml beaker.

Add 20 ml of distilled water, a magnetic stir bar and stir to dissolve the fertilizer. If it does not completely dissolve, add another 20 ml of distilled water and stir. Continue until all the fertilizer is dissolved or no more solid appears to dissolve. Do not exceed 100 ml of distilled water. If water-insoluble components remain in the beaker, vacuum filter and save the filtrate (liquid).

Determine the volume of 0.400 M $MgSO_4$ needed to precipitate the phosphate based on the label analysis and your mass of fertilizer.

Add this volume to the beaker. Calibrate the pH meter. Place the pH probe into the beaker and record the pH.

Add 6 M aqueous ammonia (NH_3) to the beaker in 5-ml increments until the pH is at least 9.0. Stand the mixture in an ice bath for 30 minutes. Place a piece of filter paper into the top of a Buchner funnel. Weigh the top of the Buchner funnel with the filter paper. Use vacuum filtration to isolate the product (see diagram at right). First, using distilled water, seal the filter paper over the holes before you filter the solution. Rinse with 2 15-ml portions of distilled water. Pull a vacuum for 5 minutes to dry the product as much as possible. The product contains excess water which must be allowed to evaporate. Remove the top of the Buchner funnel and cover with a Kimwipe. Reweigh the Buchner funnel top. Determine the mass of P_2O_5 in your fertilizer sample and its mass percentage.

9.8 Methods for Potassium Estimation

The most important potassic fertilizers are muriate of potash (MOP) and sulphate of potash (SOP). Potassium schonite and potassium syngenite can also be used potassic fertilizer. Almost all potassium are highly water soluble, like other alkali metal salts. This very property makes the analytical procedure difficult, as K is very difficult to obtain as a precipitate, which is required by the classical gravimetric and volumetric procedures.

Potassium in potassic fertilizers can be determined by the following methods:

i) Flame photometric method
ii) Volumetric sodium tetraphenyl boron (STPB) method
iii) Gravimetric tetraphenyl boron (STPB) method
iv) Gravimetric perchlorate method
v) Gravimetric chloroplatinate method
vi) Volumetric cobaltinitrite method
vii) Gravimetric cobaltinitrite method

9.8.1 Determination of K in Fertilizers and Manures by Flame Photometric Method

Principle-: potassium content in fertilizers and manures can be determined by digesting a portion of sample with di-acid or tri-acid digestion as mentioned as already in previous section, followed by estimation of K in the acid digest by flame photometric method. Water soluble K in fertilizer sample can also be determined by dissolving a portion of sample in distilled water and estimating of K using a flame photometer. Flame photometer is quite effective for determination of K in potassic fertilizers. However a adequate precaution has to be taken during dilution of fertilizer solution due to high concentration of K in fertilizer. The principle of flame photometer is that when a solution containing a metallic salt or other metallic compound is aspirated into a flame, following events occur in rapid succession i) evaporation of solvent leaving a solid residue ii) vaporization of solid residue into gaseous salt iii) dissociation of gaseous salt into its constituent neutral atoms iv) some of these neutral atoms are excited by thermal energy of the flame. The excited atoms, which are unstable, quickly emit characteristic radiation (photons) and eventually return to lower energy level (ground state). The measurement of the emitted photons or characteristic radiation forms the basis of flame photometery.

Preparation of standard curve:

Dissolve 1.909 g AR grade potassium chloride in 1 litre of distilled water. This gives 1000 ppm of standard K solution. By serial dilution prepare 0, 5, 10, 15, 20, 25, 30, and 40 ppm standard K solution.

Operation of the Instrument

Before attempting to use the flame photometer one must consult the operational manual provided by the manufacturer. The common steps to be followed are:

- Switch on the instrument
- Place the appropriate filter in position.
- Switch on the compressor to get air supply
- Adjust the sensitivity control to the minimum value

- Adjust the air control to give the pressure of 0.5 to 0.6 kg cm^{-3}.
- Turn the gas supply and light the gas at the burner
- Regulate the gas supply until the separate blue cones are formed in the burner hole.
- Set the zero by means of control knob against a blank
- Set the 100 with the highest concentration of standard solution (say, 40 ppm K)
- Repeat the above two steps till the above two setting are obtained
- Aspirate a series of standard solutions *viz.* 0. 5, 10, 15, 25, 30 and 40 ppm K
- Aspirate the unknown (fertilizer or manure) solution in the flame after suitable dilution
- Note the reading and calculate the concentration of unknown K from the calibration curve of standard solution
- For switching off the instrument close the gas supply first, run distilled water for about 2 minutes and then switch off the compressor. Turn off the knobs and switches.

Calculation

Supose, W g of fertilizer or manure sample was taken for digestion and volume was made 100 ml from which 5ml aliquot was pipette out and diluted to 100 ml. From standard curve, say concentration of K is calculated as X ppm (X µg K mL^{-1})

Then, quantity of K in 100 ml diluted solution is = 100 x X µg K)

This much of K is present in 5 ml of aliquot (fertilizer or manure solution)

Therefore, 5 ml of fertilizer or manure solution contains = 100 x X µg K)

Therefore, 100 ml of fertilizer or manure solution contains = (100 x X x 100)/5 µg K

Now, 100 ml of fertilizer or manure solution is obtained from W g of fertilizer

Therefore, W g of fertilizer or manure contains = (100 x X x 100 x 100)/5 µg K

Therefore, 100 g of fertilizer or manure contains = (100 x 100 x100).(5 x W)µg K= (100 x 100 x100)/(5 x W x 10^6) g K = X/(5 x W) g K

Therefore, the percentage of K in fertilizer or manure sample (%) = X/ (5 x W)

9.8.2 Estimation of K by Volumetric Sodium Tetraphenylboron (STPB) Method

Principle This method is applicable to both mixed and straight potassic fertilizers. In the volumetric method, potassium from the fertilizer sample is first extracted with ammonium oxalate. Potassium in solution is ten precipitated with an excess of sodium tetraphenylboron (STBP) as potassium tetraphenyborate (PTPB). The unreacted STPB is then back titrated with benzalkonium chloride (trade name Zephiron chloride) or quaternary ammonium chloride solution using clayton yellow (thiazole yellow) as indicator to estimate the quantity used for precipitation of potassium. This is basis forvolumetric analysis of potassium by STPB method.

$$Na[B(C_6H_5)_4] + K^+ \longrightarrow K[B(C_6H_5)_4] + Na^+$$

The NH4+ interferes during precipitation, and hence it is first complexed with formaldehyde (HCHO) under slightly alkaline condition before precipitation of K. It is to be noted that HCHO does not complex quaternary ammonium compounds and thus does not interfere with titration. The Cl^- and SO_4^- do not interfere during titration.

Reagents

i) 20 % NaOH solution: Dissolve 200 g of NaOH in 1 L of distilled water.

Formaldehyde (HCHO) solution (37% HCHO): AR grade reagent

iii) Sodium tetraphenylboron (STPB) solution (1.2 % approx.): dissolve 12 g STPB in approximately 800 ml of distilled water.

Add 20-25 g $Al(OH)_3$, stir for 5 minutes and filter through Whatman No. 40 0r 42 into 1 L volumetric flask.

Rinse the beaker sparingly with water and add to filtrate.

Collect entire filtrate, ad 2 ml of 20% NaOH solution, dilute to volume with water and mix thoroughly.

Let it stand for 48 h and standardize.

Adjust so that 1 ml STPB= 1%K_2O

Store at room temperature

iv) Bezalkonium chloride (Zephiron chloride) or quaternary ammonium chloride solution (approx. 0.625%): Dilute 50 ml of 12.8% Zephiron chloride to 1 L with water, mix and standardize.

v) Clayton yellow (Thiazole yellow) indicator (0.04%): Dissolve 40 mg in 100 ml of distilled water. It works at pH range from 12.2 to 13.2 and colour change is from yellow to amber.

vi) Ammonium oxalte $[(NH_4)_2C_2O_4]$ solution (4%): Dissolve 4 g ammonium oxalate in 100 ml of distilled water.

Standardization of Benzalkonium chloride (Zephiron chloride) solution

- Take 1.0 ml STPB solution in 250 ml conical flask.
- Add 20-25 ml distilled water followed by 1 ml of 20% NaOH, 2.5 ml of HCHO and 1.5 ml of 4% ammonium oxalate solution.
- Add 6-8 drops of clayton yellow indicator.
- Titrate to pink with Benzalkonium chloride solution, using 10 ml semi-micrpo burette.
- Adjust Benzalkonium chloride solution so that 2.0 ml = 1.0 ml STPB solution.

Standardization of Sodium tetraphenylboron (STPB) solution

- Dissolve 2.5 g of AR grade potassium dihydrogen orthophosphate (KH_2PO_4) (previously dried at 60 ^{0}C for 24 h) in water in 250 ml volumetric flask.
- Add 50 ml of 4% ammonium oxalate solution.
- Transfer 15 ml aliquot (51.61 mg K_2O or 43.01 mg K) in 100 ml volumetric flask.
- Add 2 ml 20% NaOH, 5 ml of HCHO and 43 ml of STPB reagent.
- Mix Dilute to volume with water and mix thoroughly.
- Allow the precipitate to stand for 5-10 minutes and pass through dry filter paper (Whatman No. 42).
- Transfer 50 ml aliquot of filtrate in the 250 ml conical flask.
- Add 6-8 drops of clayton yellow indicator.
- Titrate the excess STPB with Benzalkonium solution to pink end point.

Calculation

F= 34.61/(43.0-x) = % K_2O ml of STPB reagent.

Where, x= ml of Benzalkonium chloride solution.

This factor applies to all fertilizers if 2.5 g sample is diluted to 250 ml and 15 ml aliquot is taken for analysis. If results are to be expressed as K rather than K_2O, substitute 28.73 for 34.61 in calculating the value of F.

Preparation of Sample solution

a) **Straight potassic fertilizers (MOP, SOP, Potassium magnesium sulphate and Kainite):** Dissolve 2.5 g prepared sample and filtrate to 250 ml with distilled water. (When interfering substances such as NH_4, Ca, Al, etc. are present proceed as follow).

b) **Mixed fertiliozers (NPK fertilizers):** (The interfering substances such as NH4, Ca, Al etc have to be removed before precipitation)

  - Place 2.5 g of prepared sample in 250 ml volumetric flask.
  - Add 125 ml water followed by 50 ml of 4% $(NH_4)_2C_2O_4$ solution.
  - Add 1 ml of diglycol stearate to prevent foaming, if necessary.
  - Boil for 30 minutes, add slightly excess of NH_4OH
  - Cool and dilute to 250 ml with water, mix thoroughly and pass through dry filter paper (Whatman No.42 or equivalent).

Procedure for K determination

Pipette out 15 ml aliquot of sample solution prepared as in (a) or (b) above to 100 mL volumetric flask.

Add 2 ml of 20% NaOH followed by each of HCHO.

Add 1 ml of standard STPB solution for each 1 per cent K_2O expected in sample plus additional 8 mL excess to ensure complete precipitation.

Dilute to volume with water, mix thoroughly, and allow toprecipitate to stand for 5-10 minutes.

Filter it through dry filter paper (Whatman No. 42)

Pipette out 50 ml filtrate to 125 ml conical flask.

Add 6-8 drops of clayton yellow indicator.

Titrate excess STPB with standard benzalkonium chloride solution to pink end point.

Calculation

% K_2O = (A-B) x F

Where, F= %K_2O per ml of STPB reagent

A = ml of STPB taken

B = ml of benzalkonium chloride solution used for titration of excess STPB.

Precautions

Do not pre-mix the NaOH and HCHO solution because such mixtures are not stable and loose their capacity to complex NH_4^+

Acetone is a good solvent for the tetraphenyl boron precipitate; therefore, its use also helps in clearing glassware.

While calculating the factor % as % K_2O per ml of STPB, the figure of 34.61 is the actual percentage of K_2O present in standard KH_2PO_4.

9.9 Methods for Secondary Nutrient Estimation

9.9.1 Determination of Sulphur in Manures and Fertilizers

Principle: Most of the sulphur present in manures is in organic forms. For determination of total sulphur in manure sample it is first converted to inorganic SO_4^{2-} form following di-acid digestion (HNO_3 +$HClO_4$). Suplhate in the acid extract is then determined by gravimetric or turbidemetric method. Total S in fertilizers containing only SO_4^{2-} form can be directly determined by gravimetric or turbidimetric method, while,the non sulphate inorganic S in fertilizer like sulphideores of iron (FeS_2), copper (CuS_2), arsenic *etc.*, can be converted first into sulphate S by oxidizing with bromine in carbon tetrachloride solution, determined by gravimetric or turbidimetric method.

Digestion of organic Sulphur in manure sample

- Take 1.0 g manure sample in 150 ml conical flask and add 10-12 ml of di-acid (HNO_3: $HClO_4$) mixture.
- Place the contents of the conical flask on a hot plate and continue digestion till white fumes come for 10 minutes.
- Remove the flask from the hot plate and transfer the sample to 100 ml volumetric flask and dilute the sample to 100 ml with distilled water.

- Determine the S content in manure sample either by turbidimetric or gravimetric method as mentioned in previous section
- Digestion of Non-sulphate inorganic S in fertilizer
- Take 2.0 g fertilizer sample in a 250 ml beaker. Add 20 ml of 10% bromine in CCL_4 solution and allow it to stand for 30 minutes with intermittent mixing by swirling the beaker under a fume hood.
- Add 50 ml of conc. HNO_3 and boil it on a hot plate until all bromine is expelled.
- Cool the content and add 50 ml of water.
- Filter the content through Whatman No. 5 filter paper or equivalent into 250 ml volumetric flask.
- Wash the beaker and filter paper with warm (60 ^{0}C) water and make up volume to 250 ml mark with distilled water.
- Determine the S content in fertilizer sample either by turbidimetric or gravimetric method as mentioned in previous section

9.9.2 Estimation of S in Fertilizer and Manure and Turbidimetric Method

Principle: Solution containing sulphate (obtained after digestion of fertilizer or manure) can be determined by turbidimetric method where the intensity of turbidity (dispersed phase) caused by formation of colloidal barium sulphate suspension is measured using a spectrophotometer. The suspension (precipitate) of $BaSO_4$ is formed due to the reaction of finely ground pure $BaCl_2$ crystal with sulphate in solution. The pH of the solution is maintained at about 4.8 using a buffer solution. The stability of the suspension of $BaSO_4$ is stabilized by gum acacia, which helps to prevent rapid settling of the $BaSO_4$ precipitate. Sodium chloride and HCl are used before the addition of solid $BaCl_2$ in order to inhibit the growth of microcrystal of $BaSO_4$.

$$SO_4^{2-} + BaCl_2\,(\text{solid}) \longrightarrow BaSO_4 + Cl^-$$

Reagents

- Solid $BaCl_2$ crystal: dgind pure AR grade $BaCl_2$ crstal in a porcelain pestle and mortar and pass through 20-mesh sieve and retain them on a 60 mesh sieve.
- Sodium chloride-hydrochloric acid buffer reagent: Dissolve 60 g AR NaCl in 200 ml of distilled water. Add 5 ml of pure conc. HCl acid and dilute to 250 ml.
- Glycerol-ethanol solution: Mix glycerol and ethanol in the ratio 1:1.
- Standard solution of S (100 ppm): Dissolve 0.543 g AR grade potassium sulphate in distilled water and dilute to 1 L. This will give 100 ppm standard solution.

Procedure

- Take 0, 0.5, 1.0, 2.0, 3.0, 4.0, and 5.0 ml of 100 ppm standard solution S in a series of 25 ml of volumetric flask
- Take 5 ml of aliquot of test solution in another 25 ml of volumetric flask.
- Add 10 ml of NaCl-HCl solution and 5ml of glycerol-ethanol solution to each volumetric flask.
- Dilute the content nearly to 20 ml with distilled water.
- Add 1 mL of gum acccia solution and 1 g (approx.) of $BaCl_2$ crystal to each flask.
- Shake the flask for 1 minute until the $BaCl_2$ crystal are dissolved.
- Make up the volume to 25 ml mark with distilled water.
- Measure the intensity of the turbidity of standard solutions as well as test solution as absorbance or transmittance by a spectrophotometer at 420 nm wavelength or by a colorimeter using a blue filter.
- Draw standard curve by plotting the concentration of standard solution versus their respective absorbance and from the standard curve calculate the concentration of test solution.

Calculation

From the graph, let the conc. Of S in the test solution is X ppm (X µg S per ml).

Then, quantity of S in 25 ml suspended solution is = 25 x X µg S

This much of S is present in 5 ml of test solution

Therefore, 100 ml of test solution contains = (25 x X 100)/5 µg S

Let, 100 ml of test solution is obtained from W g of manure or fertilizer sample.

Therefore, W g of manure or fertilizer contains = (25 x X 100)/5 µg S

Therefore, 100 g of manure or fertilizer contains = (25 x X 100 x 100)/ (5 x W) µg S

= (25 x X x100x100)/(5 x W x 10^6) g S = X/(20 x W)g S

Therefore, the percentage of S in manure or fertilizer (%) = X/(20 x W)

Note/Precaution

i) The precipitation of $BaSO_4$ and the turbidity is a sensitive process and hence it is necessary to standardize the conditions, which should be strictly followed. $BaSO_4$ tends to precipitate if the solution is not properly acidified. On the other hand, if the solution is too acidic, precipitation of $BaSO_4$ is too low. Therefore, a buffer salt solution is used to maintain pH of the solution at about 4.8.

ii) Crystals of $BaCl_2$ should be added to the sulphate solution in the solid state of definite size and not as solution. The size of the crystals will determine their rate of reaction with sulphate.

9.9.3 Estimation of S in Fertilizers and Manures by Gravimetric Method

Principle: In gravimetric method, sulphate in solution (obtained after digestion of fertilizer or manures) can be precipitated by reacting with dilute solution of $BaCl_2$ as $BaSO_4$ which can be filtered off, washed thoroughly with warm distilled water, carefully ignited usually at 600 °C and weighed as $BaSO_4$.

$$MSO_4 + BaCl_2 \longrightarrow BaSO_4 \downarrow + MCl_2$$

Where M represents metal

In presence of cations like Na^+, K^+, Ca^{2+}, Li^+, Al^{3+}, Cr^{3+}, Fe^{3+} *etc.* co-precipitation of sulphates of these ions occur and cause interferences which has to be removed or masked before precipitation of sulphate with $BaCl_2$

Reagents

i) Barium chloride solution (5% $BaCl_2$). Dissolve 5 g $BaCl_2$ in 100 ml of distilled water.
ii) Concentrated HCl
iii) Concentrated HNO_3
iv) Silver nitrate (1%): Dissolve 1.0 g of $AgNO_3$ in 100 ml of distilled water
v) Bromine in carbon tetrachloride (10%): Dissolve 10 g bromine in 100 ml of CCl_4
vi) Di-acid mixture (HNO_3:$HClO_4$ in 9:4 ratio) Mix 90 ml of HNO_3 with 40 ml of $HClO_4$.

Procedure

Take 25 ml of aliquot of sulphate solution (containing 500 to 600 mg S) in a 400 ml beaker provided with stirring rod and watch glass cover

Add 0.5 ml of conc. HCl acid to it and dilute the content nearly to 200 ml with distilled water.

Heat the content in the beaker to boiling and run 10 ml of warm 5 % $BaCl_2$ solution from a burette drop by drop with a constant stirring.

Allow the precipitate to settle for a minute or two and then add few drops of $BaCl_2$ solution for testing complete precipitation.

Add 3 ml of excess $BaCl_2$ solution very slowly with constant stirring.

Allow the precipitate to settle after complete precipitation for 3 to 4 h.

Filter the precipitate through Gooch crucible using Whatman No. 40 filter paper.

Wash the precipitate with hot water 8-10 times to make the precipitate free from Cl^-. This can be done by testing the filtrate with $AgNO_3$ (1%) solution until no precipitate of AgCl results.

Transfer the precipitate along with filter paper to a previously weighed silica crucible, place the crucible in a muffle furnace and ignite to red hot at 800 °C to a constant weight.

Cool the crucible for some time and then transfer into a desiccator and take the weight of precipitate of $BaSO_4$.

Calculation

Suppose 20 g fertilizer sample is dissolved in 250 ml of distilled water, from which 25 ml of aliquot is used for precipitation.

Let, the weight of precipitate of $BaSO_4$ is W g

Now, 1 g of $BaSO_4$ contains 0.1374 g of S

Therefore W g of $BaSO_4$ contains = (0.1374 x W) g of S

This much of S comes from 25 ml of aliquot used for precipitation

Therefore 250 ml aliquot contains = (0.1374 x W x 250)/ 25 g of S

Therefore 100 g fertilizer sample contains = (0.1374 x W x 250 x100) / (25 x 2) g of S

Thus, percent S (%) = 68.7 x W

9.9.4 Determination of Ca and Mg by Complexometric Method

Principle: Calcium and magnesium in fertilizers and manure sample containing insoluble Ca and Mg can be determined by digesting the sample with di-acid mixture and estimating the Ca and Mg in the acid digest by complexometric titration with ethylene diamine tetra acid acid (EDTA) using Eriochrome Black T (EBT). While fertilizer containing soluble Ca and Mg can be titrated directly with EDTA after diluting the sample suitably. Presence of other metals like Co, Ni, Cu, Zn and Mn may interfere in the determination of Ca and Mg using Eriochrome Black T indicator. Their interference can be overcome by the use of hydroxylamine hydrochloride (which reduces some of the metals to the valency states) and also by sodium or potassium cyanide, which forms very stable complexes and act as masking agent. Alternatively, both Ca and Mg can be determined by atomic absorption spectrophotometer. The compexometric method of determination of Ca and Mg are described here.

Reagents

1. Ammonium chloride-ammonium hydroxide (NH_4Cl-NH_4OH) buffer solution (pH 10): dissolve 67.5 g NH_4Cl in 200 ml of distilled water and add 570 ml of conc. NH_4OH (sp gravity 0.88-0.90) and dilute to 1 L with distilled water.
2. EDTA solution (0.01 M): Dissolve 3.722 g of di-sodium salt of EDTA in water and dilute 1 L. Standardize the EDTA against Ca Solution (0.01N)
3. Eriochrome Black T (Solochrome Black T): Dissolve 0.4 g indicator powder in 100 ml of methyl alcohol. Alternatively, mix thoroughly 1.0 g of powder dyestuff with 100 g of NaCl and use a pinch for each titration.
4. Murexide (Ammonium salt of puprpuric acid): It can be prepared by suspending 0.5 g powder dyes tuff in 100 ml water, shaking thoroughly and allowing the un-dissolved portion to settle. The saturated liquid is used

for titration. This has to be prepared fresh. Alternatively it can be prepared by mixing with pure NaCl in the ratio of 1:500 and use approximately 0.2-0.4 g in each titration.

5. Triehtanol amine (TEA): Reagent grade
6. NaOH (10%w/v) solution: Dissolve 10 g NaOH in water and dilute to 100 ml.
7. Standard Ca Solution: Dissolve 0.5 g of AR grade $CaCO_3$ (dried at 150 ^{0}C) in 5 ml of 6 N HCl and boil to remove CO_2 and make up volume to 1 L.
8. Hydroxylamine-hydrochloride (NH_2OH-HCl) solution (5% w/v): Dissolve 5 g of hydroxylamine – hydrochloride in 100 ml of distilled water.
9. Potassium cyanide (KCN) (*Caution: poison)*: Dissolve 1 g of KCN in 100 ml of distilled water.
10. Potassium ferrocyanide [$K_4Fe(CN)_6$]: Dissolve 4 g of reagent potassium ferrocyanide trihydrate in 100 ml of distilled water.

Standardization of EDTA with standard Ca solution

- Pipette out 10 ml of standard Ca solution (0.01N) in a 100 ml conical flask or preferably the pH dish and add 25 ml of distllede water.
- Adjust the pH by adding 15 ml of NH_4Cl-NH_4OH buffer solution.
- Add 10 drops each of KCN, hydroxylamine hydroxchloride, potassium ferrocyanide and TEA solution.
- Add few drops of EBT indicator and titrate against EDTA solution till the colour changes from red to a permanent blue.

Note the volume of the EDTA solution consumed and calculate the strength of EDTA solution as follows

$N_1V_1 = N_2V_2$ or $N_1 = N_2V_2/V_1$

Where N_1 = strength of EDTA solution; V_1 = volume of the EDTA solution N_2= Strength of the standard Ca Solution; V_2 = volume of standard Ca solution

Procedure for Ca^{2+} only

- Pipette out 10 ml of aliquot in a 100 ml conical flask or preferably porcelain dish and 25 ml of distilled water.
- Adjust the pH by adding 15 ml of 10 % NaOH solution.
- Add 10 drops each of KCN, hydroxylamine hydrochloride, potassium ferrocyanide
- Add a pinch of few drops of Murexide indicator and titrate against EDTA solution till the colour changes orange red to a bluish violet.

Procedure for Ca^{2+} and Mg^{2+}

- Pipette out 10 ml of aliquot in a 100 ml conical flask or preferably porcelain dish and 25 ml of distilled water.

- Add 15 ml of NH_4Cl-NH_4OH buffer solution to it.
- Add 10 drops each of KCN, hydroxylamine hydrochloride, potassium ferrocyanide and TEA solution.
- Allow a few minutes for the reaction to take place.
- Add few drops of EBT indicator and titrate against EDTA solution till the colour changes from red to permanent blue. (**Note:** Colour changes are wine red, purple blue to permanent blue. If over titrated, it becomes green.

Calculations

For Ca^{2+} by using Murexide indicator

Let weight of the fertilizer material taken is W g and it is dissolved/diluted to 100 ml

Let the volume of standard EDTA (0.01 M) solution is V_1 ml

Now, 1 m of 0.01 M EDTA solution = 0.4008 mg of Ca

Therefore, V1 ml of 0.01 M EDTA solution = (0.4008 x V_1) mg of Ca

This much of Ca is obtained from 10 ml of aliquot

Therefore, 10 ml of aliquot contains = (0.4008 x V_1) mg of Ca

100 ml of aliquot contains = [(0.4008 x V_1 x 100)/10] mg of Ca

= [(0.4008 x V_1 x 100)/(10 x 1000)] mg of Ca

Therefore, W g fertilizer sample contains = [0.4008 x V_1 x 100)/ (10 x 1000 x W] g of Ca

100 ml aliquot contains = [(0.4008 x V_1 x 100 x 100)/(10 x 1000 x W)] g of Ca

=[(0.4008 x V_1)/W] g of Ca

Therefore, per cent Ca in fertilizer sample (%) = 0.4008 x V_1/W

For Ca^{2+} and Mg^{2+} using EBT indicator

Let weight of the fertilizer material taken is W g and it is dissolved/diluted to 100 ml

Let the volume of standard EDTA (0.01 M) solution consumed for both Ca^{2+} and Mg^{2+} is V_2 ml

Therefore, the amount of EDTA (0.01 M) consumed for Mg^{2+} is = (V_2-V_1) ml

Now, 1 ml of 0.01 M EDTA solution = 0.2432 mg of Mg

Therefore, (V_2-V_1) ml of 0.01 M EDTA M solution = [0.2432 x (V_2-V_1)] mg of Mg^{2+}

This much of Mg is obtained from 10 ml of aliquot

Therefore 10 ml of aliquot contains = [0.2432 x (V_2-V_1)] mg of Mg^{2+}

100 ml aliquot contains = [0.2432 x (V_2-V_1) x100 /10] mg of Mg^{2+}

= [0.2432 x (V_2-V_1) x100 /(10 x 1000)] g ofMg^{2+}

Therefore W g sample contains = [0.2432 x (V_2-V_1) x100 / (10 x 1000 x W)] g of Mg^{2+}

100 g sample contains = [0.2432 x (V_2-V_1) x100 x 100 /(10 x 1000 x W)] g of Mg^{2+}

= [0.2432 x (V_2-V_1)

Therefore, percent Mg in fertilizer sample (%) = 0.2432 x (V_2-V_1) /W

9.9.5 Methods for Determination for Micro-nutrients Estimation

Micronutrients in manure sample are present in organic form, which can be determined by digesting the sample with di-acid mixture and estimating their content in the acid digest either by volumetric, gravimetric, spectrophtometric or atomic absorption spectrophotometric method. These methods are also used for determination of micro-nutrients in straight fertilizers, while only AAS method is employed for estimation of micronutrient in fertilizers mixtures because of their low concentration and chances of interferences from other elements. Volumetric and gravimetric methods are classical in nature, but these are time consuming, laborious and tedious in practice. Colorometric and spectrophotometric methods are faster than volumetric and gravimetric methods, yet comparable in precision provided necessary precaution are taken during sample preparation and subsequent analysis. The method employed in AAS is fast, precise, easy and suitable for determination of all micro-nutrients, though cost of instrument and recurring expense are very compared to the other methods.

Determination of micronutrients by AAS method

Principle: Most of the micronutrient fertilizers are soluble in water. A known quantity of fertilizer sample is dissolved in distilled water and diluted to working range suitable for that particular micronutrient to be analysed. A series of standard solution is prepared for constructing a standard curve for element. The standard solution and sample are aspirated into flame of AAS using suitable fuel and oxidant and measured the absorbance at a wavelength specific to the metal under analysis. A specific hollow cathode lamp is used for each element. The concentration of the unknown test solution is then worked out from the standard curve.

Reagents

i) Concentrated HCl

ii) All-glass double distilled water.

iii) Acidified water (pH 2.5± 0.5): Dissolve 10 ml of 10% H_2SO_4 in 10 L of double distilled water and adjust water pH at 2.5 with the help of NaOH or H_2SO_4

iv) Standard micro-nutrients solution: Prepare standard solution of each of the micro-nutrients. This can be done as per the procedure given below.

Standard Zn (1000 ppm) Solution: Dissolve 1.0 g of pure Zinc metal in 30 ml of 6 N HCl, heat the content and dilute to 1 L with acidified water. A working solution of 0.2 to 2.0 ppm is then prepared by serial dilution with acidified water.

Standard Cu (1000 ppm) solution: Dissolve 1.0g of pure Cu metal in 30 ml of 6 N HCl, heat the content and dilute to 1 L with acidified water. A working solution of 0.2 to 1.0 ppm is then prepared by serial dilution with acidified water.

Standard Fe (1000 ppm) solution: Dissolve 1.0g of pure Fe metal in 30 ml of 6 N HCl, heat the content and dilute to 1 L with acidified water. A working solution of 2 to 20 ppm is then prepared by serial dilution with acidified water.

Standard Mn (1000 ppm) solution: Dissolve 1.0g of pure Mn metal in 30 ml of 6 N HCl, heat the content and dilute to 1 L with acidified water. A working solution of 0.5 to 5.0 ppm is then prepared by serial dilution with acidified water.

Standard Mo (1000 ppm) solution: Dissolve 1.5g of pure MoO_3 metal in 1 L of acidified water and make up the volume to 1 L mark. A working solution of 20 to 60 ppm is then prepared by serial dilution with acidified water.

Standard B (5000 ppm) solution: Dissolve 44.095 g of pure sodium borate ($Na_2B_4O_7.10H_2O$) in 1 L of warm water and make up the volume to 1 L mark. A working solution of 400to 1600 ppm is then prepared by serial dilution with acidified water.

Preparation of different micro-nutrient fertilizer sample

Dissolve required quantity of fertilizer sample in water to get approximately 1000 ppm of element to be determined. It is then diluted to the working range of the specific element with the acidified water. Some micro-nutrient fertilizers, the quantity to be dissolved to get approximately 1000 ppm solution and further dilution to be done with acidified water are given in appendices.

Procedure

Prepare a series of standard solution from the working range of standard of particular element to be analysed.

Aspirate the standard solution in the AASD at a specific wavelength of element to be analysed using a hollow cathode lamp of that element.

Draw a standard curve using concentration (ppm) of standard solution versus corresponding absorbance.

Aspirate the test solution after suitable dilution and note down the absorbance.

From the standard curve calculate the concentration of the test solution.

Calculation

Let, W g fertilizer sample is dissolved in 1000 ml water. Out of which suppose, 10 ml aliquot is further diluted to 500 ml and concentration is measured in AAS.

Suppose, the concentration measured from standard curve is X ppm (µg me^{-1})

Therefore, 500 ml of diluted fertilizer sample contains – (500 x X) µg of element

This much element is present in 10 ml of undiluted fertilizer sample

Therefore, 1000 ml undiluted fertilizer sample contains = (500 x X x 1000/10)µg

Therefore, W g of fertilizer sample contains = [(500 x X x 1000)/ (10 x W) µg

Therefore, 100 g of fertilizer sample contains = [(500 x X x 1000 x 100)/ (10 x W) µg

= [(500 x X x 1000 x 100)/(10 x Wx 10^6)] g of element

= (5 x X/W) g of element

Therefore, per cent element present in fertilizer sample (%) = (5 x X/W)

This can be calculated directly as :

Percentage of element (%) = [500 x X x 100 x 100)/ (10 x W x 10^6)

=[X x (500 x 1000/10)/ (W x 10^4)]

= [X x dilution factor/ (W x 10^4)]

$$\text{Thus, percentage of element (\%)} = \frac{\text{X ppm from graph x dilution factor}}{\text{Weight of sample (g)} \times 10^4}$$

CHAPTER-10

Water Analysis

Knowledge of irrigation water quality is critical to understanding management for long-term productivity. Irrigation water always contains some soluble salts irrespective of its source. The suitability of waters for a specific purpose depends on the types and amounts of dissolved salts. Some of the dissolved salts or other constituents may be useful for crops. However, the quality or suitability of waters for irrigation purposes is assessed in terms of the presence of undesirable constituents, and only in limited situations is irrigation water assessed as a source of plant nutrients. Irrigation water quality is evaluated based upon total salt content, sodium and specific ion toxicities. Some of the dissolved ions, such as NO_3^-, are useful for crops.

There are different limits of purity established for drinking water, water to be used for industrial purposes and for agriculture. Therefore, it may be possible that water which is not fit for drinking and industrial use may be quite suitable for irrigation. This publication deals with irrigation waters only. The most important characteristic that determine the quality of irrigation waters are:

1. pH
2. Total concentration of soluble salts judged through electrical conductivity (EC).
3. Relative proportion of sodium to other cations such as Ca and Mg referred as Sodium Adsorption Ratio (SAR).
4. Concentration of boron or other elements that may be toxic to plants
5. Concentration of carbonates and bi-carbonate as related to the concentration of calcium plus magnesium referred as Residual Sodium Carbonate (RSC).
6. Content of anions such as chloride, sulphate and nitrate.

The standards fixed for above each aspect act as an index to describe the analytical data on the above parameters. Water coming from industries as effluents and domestic wastewater as sewage may contain some specific plant nutrients. These waters may be useful for irrigation of field crops if assessed for toxic/pollutant metals and organic and microbial constituents with regard to their suitability or otherwise. Determination of organic constituents is generally carried out in two categories: (i) organic substances that quantify an aggregate amount of organic C; and (ii) individual or specific organic substances, such as benzene, DDT, methane, phenol and endosulphan. Important determinations are the chemical oxygen demand (COD), which gives the total organic substances, and the biochemical oxygen demand (BOD), which gives the amount of total biodegradable organic substances in the water sample.

10.1 Important Indices of Practical Utility to Judge Irrigation Water Quality

The following standards are laid down:

10.1.1 Electrical Conductivity (EC)

The concentration of total salt content in irrigation waters, estimated in terms of EC, is the most important parameter for assessing the suitability of irrigation waters. Generally, all irrigation waters with an EC of less than 2.25 mS/cm are considered suitable except in some unusual situations, e.g. very sensitive crops and highly clayey soils of poor permeability. The ideal value is less than 0.75 mS/cm (Richards, 1954).

10.1.2 Sodium Adsorption Ratio (SAR)

The SAR is calculated in order to determine the sodicity or alkalinity hazard of irrigation waters:

$$SAR = \frac{Na^{+}}{\frac{Ca^{2+} + Mg^{2+}}{2}}$$

Where, the concentration of cations is in meq/litre.

Based on the value of SAR, waters can be rated into different categories of sodicity as under Richards, 1954).

Safe	< 10
Moderately Safe	10-18
Moderately unsafe	19-26
Unsafe	> 26

10.1.3 Residual Sodium Carbonate (RSC)

This index is important for carbonate and bicarbonate rich irrigation waters. It indicates their tendency to precipitate calcium as $CaCO_3$. RSC is calculated as below.

RSC (meq/litre) = $(CO_3^{2-} + HCO_3^-) - (Ca^{2+} + Mg^{2+})$

Concentrations of both cations and anions are in meq/litre. Sodicity hazard in terms of RSC is categorized as under (Richards, 1954).

Safe	<1.25
Moderate	1.25 – 2.5
Unsafe	>2.5

The limits can vary depending upon types of soils, rainfall and climatic conditions. Higher RSC values can be considered safe for sandy soils in high rainfall area (> 600 mm/annum).

10.1.4 Mg/Ca Ratio

It is widely reported that calcium and magnesium do not behave identically in soil system and magnesium deteriorates soil structure particularly when irrigation water is sodium dominated and highly saline. High level of Mg usually promotes higher development of exchangeable Na in irrigated soils. Based on ratio of Mg to Ca, waters are categorized as below.

Safe	<1.5
Moderate	1.5–3.0
Unsafe	>3.0

10.1.5 Boron Content

Boron, an essential micronutrient is required in very small quantity by the crops. It becomes toxic, if present beyond a particular level. In relation to boron toxicity, water quality ratings are as below.

Low hazard	<1 µg B/ml
Medium hazard	1-2 µg B/ml
High hazard	2-4 µg B/ml
Very high hazard	>4 µg B/ml

Each of the above parameters has bearing on the quality of irrigation water. However, each water source will have its specific suitability or hazardous nature depending upon the presence (and the degree) or absence of each of the constituents. Different chemical constituents interact with each other and cause a complex effect on soil properties and plant growth. The waters with low SAR and low EC are widely suitable but when a value of any one of these parameters or both increases in its content, the waters become less and less suitable for irrigation purposes. The selection of crops for such situation becomes critical. Salt tolerant crops can be grown in such areas. Soil type is also an important consideration under such situations.

Upper permissible limits of EC, SAR, RSC and B are indicated below (Table 10.1) for soil having varying amounts of clay and for growing tolerant (T) and semi-tolerant (ST) crops.

Table 10.1 : Suitability of irrigation water for semi-tolerant and tolerant crops in different soil types

Textural category	Upper permissible limit							
	EC (dS/m)		SAR		RSC (meq/litre)		B (μg/ml)	
	ST	T	ST	T	ST	T	ST	T
Above 30% Clay	1.5	2.0	10	15	2	3	2	3
20-30% Clay	4.0	6.0	15	20	3	4	2	3
10-20% Clay	6.0	8.0	20	25	4	5	2	3
Below 10% Clay	8.0	10.0	25	30	5	6	3	4

10.1.6 Trace Elements

Presence of trace elements or heavy metals cause reduction in crop growth, if their concentration increases beyond a certain level in irrigation waters and if such waters are used continuously. However, such elements are normally not a problem in common irrigation waters. They can be of concern when industrial effluent water is used for irrigation.

10.2 Methods of Water Sample Collection

A representative sample (500 ml) is collected in glass or polyethylene bottle which should be properly washed/rinsed with the same water which is being sampled. The floating debris or any other contaminant should be avoided while collecting the samples. After proper labeling, such as source of water, date of collection and the type of analysis required, the sample should be sent to the laboratory without undue delay.Some of the anions like SO_4 and NO_3 may be quite low in irrigation waters. Hence, large volume of sample has to be first concentrated by evaporating to about 100 ml to obtain their detectable amounts.

10.3 Analytical Methods

10.3.1 pH

The pH is determined by taking about 50 ml of water sample in 100 ml clean beaker by using pH meter as described under previous chapter, soil analysis

10.3.2 Electrical Conductivity (EC)

The conductivity meter cell is filled with water sample and the EC is determined as described under chapter 3, soil analysis. The results are expressed as mS/cm.

10.3.3 Calcium and Magnesium

The usual method for determination of Ca+Mg is by versenate (EDTA) titration (Cheng and Bray, 1951). The method for their estimation in soils is described in previous section of the chapter on soil analysis. In case of soils, generally exchangeable calcium and magnesium are estimated. Estimation of Ca + Mg in water by EDTA method is described as under:

Apparatus

- Porcelain dish
- Volumetric flasks
- Burette

Reagents

1. Standard versenate solution (EDTA): An approximately 0.01N solution of ethylene di amine tetraacetic acid disodium salt (versenate) is prepared by dissolving 2.0 g in distilled water to which 0.05 g of magnesium chloride ($MgCl_2.6H_2O$) is added and diluted to one litre. This is to be standardized against 0.01N calcium chloride solution prepared by weighing 0.500 g AR grade $CaCO_3$ (oven dried) and dissolving in minimum excess of dilute HCl (AR) followed by making up the volume to one litre with distilled water.
2. Ammonium chloride-ammonium hydroxide buffer (pH 10): 67.5 g pure ammonium chloride dissolved in 570 ml of concentrated ammonium hydroxide and made to one litre. pH adjusted to 10.
3. Eriochrome black T indicator: 0.5 g of Eriochrome black T and 4.5 g of hydroxylamine hydrochloride (AR) dissolved in 100 ml of 95% ethyl alcohol.

Procedure

1. Take 5 ml of the water sample in a porcelain dish (8 cm diameter).
2. Dilute to about 25 ml with distilled water.
3. Add 1 ml. of ammonium chloride-hydroxide buffer and 3 to 4 drops of Eriochrome black T indicator.
4. Titrated with the standard versenate solution. The colour change is from wine red to bright blue or bluish green. At the end point no tinge of the red colour should remain.

Calculation

From the volume of 0.01N EDTA (standardized against 0.01N $CaCl_2$) solution required for titration, the concentration of Ca+Mg is directly obtained in me/litre as follows:

$$\text{Ca + Mg (meq/litre)} = \frac{\text{ml versenate (EDTA) used} \times \text{normality of EDTA} \times 1000}{\text{ml aliquot taken}}$$

$$\text{Ca + Mg (g/litre)} = \frac{\text{Ca+Mg in meq/litre} \times \text{equivalent wt.}}{1000}$$

$$= \frac{\text{Ca + Mg in meq/litre} \times 32.196}{1000}$$

OR

$$\text{Ca + Mg (g/litre)} = \frac{\text{ml versenate (EDTA) used} \times \text{normality of EDTA} \times 1000 \times \text{eqt. wt.}}{\text{ml aliquot taken}}$$

$$\frac{\text{ml versenate (EDTA) used} \times \text{normality of EDTA} \times 1000 \times 32.196}{\text{ml aliquot taken}}$$

10.3.4 Sodium

Small amount of sodium is generally present even in the best quality of irrigation water. The concentration of sodium may be quite high in saline water with EC greater than 1 mS/cm and containing relatively less amount of Ca and Mg. Obviously, its estimation is of interest when the water sample tests saline (i.e. having EC above 1.0 mS/cm at 25^{0}C). The determination of Na is carried out directly with the help of flame photometer using appropriate filters and standard curves prepared by taking known concentration of Na.

Apparatus

- Flame photometer
- Volumetric flasks
- Beakers

Reagents

- NaCl (AR grade)

Procedure

1. Preparation of standard curve:
   - Take 2.5413 g of NaCl (AR), dissolve in water to make to the volume to 1 litre and this will give a solution of 1000 µg Na/ml. From this solution take 100 ml and dilute to 1 litre to obtain 100 µg Na/ml as stock solution.
   - For preparing working standards, take 5, 10, 15 and 20 ml of stock solution in 100 ml volumetric flask and make up the volume. It would give 5, 10, 15 and 20 µg Na/ml.
   - Feed the standards on the flamephotometer one by one to obtain a standard curve taking absorbance on Y-axis and respective concentrations of Na on X- axis.
2. Water samples are fed on the flamephotometer and absorbance is recorded for each sample.
3. Concentration of Na is observed against each absorbance which is in µg Na/ml.

Calculation

$$\text{Content of Na in mg/litre of water} = \frac{A \times 1000}{1000} = A$$

where,

A = absorbance reading (µg/ml) from the standard curve

Note:

- If a water sample is diluted for estimation, the quantity of sodium as observed on standard curve is multiplied by the dilution factor.
- If the water sample is concentrated before estimation, the quantity noted from the standard curve is divided by the concentration factor.
- Normally, no dilution and conc. is required.

10.3.5 Carbonates and Bi-carbonates (Richards, 1954)

The estimation is based on simple acidimetric titration using different indicators which work in alkaline pH range (above 8.2) or in acidic pH range (below 6.0).

Apparatus

- Porcelain dish
- Burette

Reagents

- Phenolphthalein indicator: 0.25% solution in 60% ethyl alcohol
- Methyl orange indicator: 0.5% solution in 95% alcohol
- Standard sulphuric acid (0.01M).

Procedure

1. Take 5 ml of the water sample (containing not more than one milliequivalent of carbonate plus bicarbonate) in a porcelain dish.
2. Dilute with distilled water to about 25 ml.
3. A pink colour produced with a few (2 to 3) drops of phenolphthalein indicates presence of carbonate and it is titrated with 0.01M sulphuric acid until the colour just disappears (phenolphthalein end point) because of alkali carbonate having been converted to bicarbonate. This is called half neutralization stage. This burette reading (volume used) is designated as Y.
4. To the colourless solution from this titration (or to the original sample of water if there was no colour with phenolphthalein) add 1 to 2 drops of methyl orange indicator and continue titration with brisk stirring to the methyl orange end point (yellow) and the final reading (volume used) is designated as Z.

Calculation

$$\text{Carbonates (meq/litre)} = 2(\text{Vol. } H_2SO_4) \text{ x Mol. } H_2SO_4 \times \frac{1000}{\text{ml of aliquot}}$$

$$= 2Y \times 0.01 \times \frac{1000}{5} = 2Y \times 2 = 4Y$$

$$\text{Carbonates (g/litre)} = \frac{2(\text{Vol. of } H_2SO_4) \times \text{Molarity} \times 1000x \text{Eq. wt. of } CO_3(30)}{\text{ml of sample } \times 1000}$$

$$= 2Y \times 0.01 \times 30 = \frac{0.12Y}{5}$$

Note:

The volume of acid used for half-neutralization of carbonate is Y, hence for full neutralization it has been assumed as 2Y.

$$\text{Bicarbonates(meq/litre)} = (\text{Z-2Y}) \times \text{molarity of } H_2SO_4 \times \frac{1000}{\text{ml of aliquot}}$$

$$= \frac{(\text{Z-2Y}) \times 0.01 \times 1000}{5}$$

$$= (\text{Z-2Y}) \text{ x } 2$$

Where carbonate is absent: Z x 2

10.3.6 Residual Sodium Carbonate (RSC)

This is an important character for assessing the suitability of irrigation water in consideration of likely sodium hazard. It is calculated from the analysis data for carbonates, bicarbonates and calcium plus magnesium in the following manner:

RSC (meq/litre) = $(Co_3^{2-} + HCo_3^{-})-(Ca^{2+} + Mg^{2+})$

Note: all expressed in meq/litre.

10.3.7 Boron

The method for Boron estimation is same as described for soils in chapter 3. The determination is carried out by azomethione-H colorimetric method. It can also be estimation on AAS. Suitable quantities of the sample may be taken depending upon the Boron content in the waters.

10.3.8 Chlorides

Mohr's titration method is most commonly used for chloride estimation. It depends upon the formation of a sparingly soluble brick-red silver chromate ($AgCrO_4$) precipitate at the end point when the sample is titrated against standard silver nitrate ($AgNO_3$) solution in the presence of potassium chromate (K_2CrO_4) as indicator. Initially the Cl^- ions are precipitated as AgCl and dark brick-red precipitate of Ag_2CrO_4 starts just after the precipitation of AgCl is over.

Apparatus

- Beakers/porcelain dish
- Burette

Reagents

1. Potassium chromate (K_2CrO_4) indicator (5%) solution: Dissolve 5 g of K_2CrO_4 in about 75 ml distilled water and add drop by drop saturated solution of $AgNO_3$ until a slight permanent red precipitate is formed. Filter and dilute to 100 ml. With high purity analytical reagent, the indicator solution can be prepared directly.
2. Standard silver nitrate solution (0.05M): Dissolve 8.494 g of silver nitrate ($AgNO_3$) in distilled water and make the volume to one litre. Standardize it against standard NaCl solution and keep in amber coloured bottle away from light.

Procedure

1. Take 5 ml of the sample in a 100 ml beaker or a porcelain dish and diluted to about 25 ml with distilled water.
2. Add 5-6 drops of K_2CrO_4 indicator (making it dark yellow), and titrate against the standard $AgNO_3$ solution with continuous stirring till the first brick-red tinge appears.
3. Run a blank to avoid error due to any impurity in chemicals.

Calculation

$$\text{Cl mg/litre of water} = \frac{X \times 1.1775 \times 1000}{\text{ml of sample}}$$

Where,

ml of water sample taken = 5

X = ml of 0.05M $AgNO_3$ consumed in titration

1.775 = factor representing mg of Cl in aliquot/sample as calculated below:

1 ml of 1M $AgNO_3$ = 1 meq of Cl

1 ml of 0.05M $AgNO_3$ = 0.05 meq of Cl = 35.5 x 0.05

= 1.775 mg of Cl (in aliquot)

10.3.9 Sulphate

While traces of sulphate occur universally in all types of waters, its content may be appreciably high in several saline waters showing EC greater than 1 dS/m at 25 ^{0}C. Sulphate can be determined gravimetrically, colorimetrically, turbidimetrically or titrimetrically. Here, the turbidimetric method is described:

Sulphate content is determined by the extent of turbidity created by precipitated colloidal barium sulphate suspension. Barium chloride solid crystals are added to ensure fine and stable suspension of $BaSO_4$ at a pH of about 4.8. It also eliminates the interference from phosphate and silicate. Fine suspension of $BaSO_4$ is stabilized by gum acacia and the degree of turbidity measured by turbidity meter or estimated by spectrophotometrically at 440 nm.

Apparatus

- Spectrophotometer
- Beakers
- Volumetric flasks

Reagents

1. Sodium acetate-acetic acid buffer: Dissolve 100 g of pure sodium acetate in 200 ml of distilled water. Add 31 ml of glacial acetic acid and make the volume to one litre. Adjust pH at 4.8.
2. Gum acacia: Dissolve 2.5 g of gum acacia in one litre of distilled water. Keep overnight and filter.
3. Barium chloride: Pure $BaCl_2$ crystals ground to pass through 0.5 mm sieve but retained on a 0.25 mm sieve.
4. Potassium sulphate solution: To make a stock solution of 10 meq S/litre, weigh 1.74g of pure K_2SO_4 salt and dissolve in one litre water.

Procedure

1. Take 5 ml of the water sample (having <1 meq S/litre) in 25 ml of volumetric flask. If the EC of water is >1 dS/m, dilute it with distilled water to bring EC below 1 dS/m.
2. Add 10 ml of sodium acetate-acetic acid buffer to maintain the pH around 4.8.
3. Add 1 ml of gum acacia and 1 g of $BaCl_2$ crystals and shake well.
4. Make the volume to 25 ml with distilled water.
5. Invert the flask several times and measure the turbidity with a spectrophotometer at 440 nm using blue filter.
6. Preparation of the standard curve: For 0, 1, 2, 3, 4 and 5 meq S/litre, pipette 2.5, 5, 7.5, 10 and 12.5 ml from stock solution containing 10 meq S/litre into 25 ml volumetric flasks. Then develop the turbidity and measure its intensity as in case of samples. Draw a curve showing sulphur concentration on x-axis and absorbance on y-axis.

Calculation

Calculate the S content of samples using the standard curve taking in to the dilution factor of 5 (5 ml made to 25 ml) expressed as meq S/litre of water.

10.3.10. Nitrate Nitrogen (NO_3^- – N)

This method depends upon the reduction of nitrate to ammonia by adding Devarda's alloy and alkali. The nitrites (NO^{2-}) if present in the sample are also reduced and determined along with NO_3^-–N.

Apparatus

- Kjeldahl distillation assembly
- Electric muffle furnace
- Desiccator

Reagents

1. Magnesium oxide (MgO): Heat the MgO at 65 ^{0}C for 2 hours in an electric muffle furnace to remove traces of $MgCO_3$ which may be present. Cool in a desiccator over solid KCl and store in tightly stoppered bottle.
2. Boric acid with mixed indicator: Weigh 20 g of boric acid and add approximately 900 ml of hot distilled water. Cool and add 20 ml of mixed indicator and make up the volume to 1 litre.
3. Mixed indicator: Dissolve 0.066 g of methyl red and 0.099 g of bromocresol green in 100 ml of alcohol.
4. Standard sulphuric acid (0.02M).
5. Devarda's alloy: Mix Cu:Al:Zn in the ratio of 50:45:5 and grind to pass through 0.15 mm sieve.

Procedure

1. Take 50 ml of water sample in the distillation flask.
2. Add 0.5 g of MgO and 0.2 g of Devarda's alloy.
3. Put the heaters on and collect the NH_4 (NO_3 coverted in NH_4 by reducing agent – Devarda's alloy) into boric acid (20 ml) having mixed indicator into conical flask, which is connected with distillation apparatus.
4. Continue distillation to collect about 35-40 ml.
5. Remove the distillate first and then switch off the heating system.
6. Titrate the distillate against 0.02M H_2SO_4 till the pink colour appears.
7. Carry out a blank simultaneously.

Calculation

$$NO_3^- \text{–N (mg/litre)} = \frac{(X\text{-}Y) \times 0.28 \times 1000}{50 \text{ (ml of sample)}} = X - Y \times 0.56$$

Where,

X = volume (ml) of 0.02M H_2SO_4 consumed in sample titration.

Y = volume (ml) of 0.02M H_2SO_4 consumed in blank titration.

0.28 = Factor (1 lit 1M H_2SO_4 = 14 g N.

Therefore,

$$1 \text{ ml } 0.02\text{M } H_2SO_4 = \frac{14 \times 0.02 \times 1000}{1000} \text{ mg N=0.28 mg N}$$

Appendices

Appendix-I : pH and quantity of lime required to reduce soil acidity (tonnes/ha)

(pH afterbuffer)	$CaCO_3$	$Ca(OH)_2$	Marl	Limestone	Dolomite
6.5	6.00	4.68	7.20	9.00	6.54
6.4	7.20	5.62	8.64	10.80	7.85
6.3	8.40	6.55	10.08	12.60	9.16
6.2	9.60	7.49	11.52	14.40	10.46
6.1	10.80	8.42	12.96	16.20	11.77
6.00	12.00	9.36	14.40	18.00	13.08
5.9	13.20	10.30	15.84	19.80	14.39
. 58	14.40	11.23	17.28	21.60	15.70
5.7	15.60	12.17	18.72	23.40	17.00
5.6	16.80	13.10	20.16	25.20	18.31

Appendix-II: General interpretation of EC values

Soil	EC (mS/cm)	Total Salt Content (%)	Crop reaction
1. Salt free	0-2	<0.15	Salinity effect negligible, except for more sensitive crops
2. Slightly saline	2-8	0.15-0.35	Yield of many crops restricted
3. Moderately saline	8-15	0.35-0.65	Only tolerant crops yieldsatisfactorily
4. Highly saline	>15	>0.65	Only very tolerant crops yieldsatisfactorily

Appendix-III: Critical limits for DTPA extractable micronutrients

	Micronutrient (µg/g soil)			
Availability	Zn	Cu	Fe	Mn
Very low	0-0.5	0-0.1	0-2	0-0.5
Low	0.5-1	0.1-0.3	2-4	0.5-1.2
Medium	1-1.3	0.3-0.8	4-6	1.2-3.5
High	3-5	0.8-3	6-10	3.5-6
Very high	>5	>3	>10	>6

Appendix-IV: The specifications of relevant hollow cathode lamps used in AAS

Specifications	Zn	Cu	Fe	Mn
Lamp current (m A°)	5	3	7	5
Wave length (nm)	213.8	324.9	248.7	279.5
Linear range (mg/l)	0.4-1.5	1.0-5.0	2.0-9.0	1.0-3.6
Slit width (nm)	0-2	0-2	0-2	0-2
Integration time (sec)	2.0	2.0	2.0	2.0
Flame		Air Acetylene		

Appendix-V: Wavelengths and corresponding colour ranges

Wavelength (nm)	Hue	Complementary Hue
<400	Ultraviolet	
400-435	Violet	Yellow green
435-480	Blue	Yellow
480-490	Greenish blue	Orange
490-500	Bluish green	Red
500-560	Green	Purple
560-580	Yellowish green	Violet
585-595	Yellow	Blue
595-610	Orange	Greenish Blue
610-750	Red	Bluish Green
>760	Infra-red	

Appendix-VI: The quantities of chemical required to make 1 litre standard solution of 100 μg/ml for different elements from their respective salts

Element	Conc. of stock solution (μg/ml)	Salt to be used	Quantity of salt required/litre
Zn	100	Zinc Sulphate($ZnSO_4.7H_2O$)	0.4398
Cu	100	Copper Sulphate($CuSO_4.5H_2O$)	0.3928
Fe	100	Ferrous Sulphate($FeSO_4.7H_2O$) or Ferrous Ammonium Sulphate	0.49640. 7028
Mn	100	Manganese Sulphate($MnSO_4.H_2O$)	0.3075

Appendix-VII: Preparation of different micronutrient fertilizer sample

Micro-nutrient fertilizer	Micro-nutrient content (%)	Amount of fertilizer to be dissolved in 1 L	Further with acidified water
$ZnSO_4.7H_2O$	21.0	0.25 g	50 times
$ZnSO_4.H_2O$	33.0	0.25 g	50 times
$FeSO_4.7H_2O$	19.0	1.00 g	20 times
Chelated-Fe	12.0	1.00 g	20 times
$MnSO_4.H_2O$	30.5	0.40 g	50 times
$CuSO_4.10H_2O$	24.0	0.25 g	100 times
$Na_2B_4O_7.10H_20$	10.5	10.0 g	No dilution
$[NH_4)_6MoO_{24}.2H_2O]$	52.0	1.0g	10 times